KEEPING *the*
BEES

KEEPING *the*
BEES

WHY ALL BEES ARE AT RISK
AND WHAT WE CAN DO TO SAVE THEM

LAURENCE PACKER

HARPER
PERENNIAL

Keeping the Bees

Copyright © 2010 by Laurence Packer.

First published in Canada by HarperCollins Publishers Ltd
in a hardcover edition: 2010
This Harper Perennial trade paperback edition: 2014

HarperCollins books may be purchased for educational, business,
or sales promotional use through our Special Markets Department.

HarperCollins Publishers Ltd
2 Bloor Street East, 20th Floor
Toronto, Ontario, Canada
M4W 1A8

www.harpercollins.ca

Library and Archives Canada Cataloguing in Publication
information is available upon request

ISBN 978-0-06-230646-3

Printed and bound in the United States

RRD 9 8 7 6 5 4 3 2

*This book is dedicated, with love and appreciation,
to my parents, Denis and Lilian, who in different ways
encouraged my interest in insects.*

CONTENTS

1 BUZZ FREE: A WORLD WITHOUT BEES
Stuck Under a Truck in the Atacama Desert, Chile   1

2 THE FUTURE OF OUR FOOD
Serenading the Bees on Cape Breton Island, Nova Scotia   17

3 HONEY, QUEENS, HARD-WORKING WORKERS AND STINGS:
MISCONCEPTIONS ABOUT BEES
Stung by Bee Killers on the Isle of Wight, United Kingdom   35

4 A BEE OR NOT A BEE? A DIFFICULT QUESTION
TO ANSWER
Insulting the Experts in Portal, Arizona   51

5 TWO BEES OR NOT TWO BEES? AN EVEN *MORE* DIFFICULT
QUESTION TO ANSWER
Uncovering Irregularities at the National History Museum, London   63

6 IT'S A BEE'S LIFE
Up Before Dawn in the Australian Outback   79

7 THE SOCIABLE BEE
Digging Nests After Dark in Calgary, Alberta   101

8 SEX AND DEATH IN BEES
Choosing a Mate in Subtropical Florida   117

9 WHERE THE BEE SUCKS, THERE HUNT I
Painful Bee Sampling in the Tehuacan Desert, Mexico   131

10 ANTI-BEES
Sexually Transmitted Child-eating Female Impersonators on a
California Sand Dune   157

11 WHAT ARE WE DOING TO THE BEES?
Bee-Free Day in Germany   173

12 THE PROVERBIAL CANARIES IN THE COAL MINE
Upsetting Ornithologists in Rome   195

13 HELP THE BEES
Dodging Hippos at the African Pollinator Summit   211

EPILOGUE
Bee Worship on the Yucatan Peninsula, Mexico   223

Acknowledgments   229
Appendix 1: Bee Families   233
Appendix 2: Bee Names   235
Sources   241
Index   263

KEEPING *the*
BEES

1

BUZZ FREE: A WORLD WITHOUT BEES

STUCK UNDER A TRUCK IN THE ATACAMA DESERT, CHILE

It was three in the afternoon and over 30°C, yet despite the heat there was not a drop of sweat on my body. The air was so dry that any perspiration was sucked from my pores before I even felt it. I could see for miles in all directions, but there was no sign of human habitation. I hadn't seen a soul for hours. The entire vista was in two colours—the ground was beige and the sky blue—and even the Andes Mountains, faintly visible on the horizon, looked blue from this distance. I had been stuck here in the Atacama Desert in northern Chile for three hours, trying to dig the back wheels of a half-ton truck out of the sand.

The day had begun well. I had found a tiny trickle of water emerging from an embankment at the side of a dirt road; there was some vegetation growing around the moisture, and some bees were visiting the flowers. I collected a few of them and then drove off towards my next sample site. Two hours of driving and three hours of digging later, and I had not seen a single living thing—no other drivers, not an insect, not a plant.

The Atacama Desert is both the driest and the oldest desert in the world. People claim that there are parts of this desert where rain has never fallen, although geologists have told me that it must have rained at least once in the past hundred years over most of this enormous stretch of barren land. There are indications of past precipitation on the surfaces of the mine tailings, extrusions from the nitrate mines that dotted the landscape with human activity a century ago. This is an eerie place to travel, as in some areas the only signs of past human habitation are the graveyards that house the dead. The mausoleums shelter the desiccated remains of the mine managers and their families (the miners themselves were given less prestigious burials). Apart from some vandalism, the coffins and their contents are exactly as they were in the early years of the twentieth century. Where it (almost) never rains, the dead become mummified.

Imagine living in a place where it rains perhaps once every hundred years. Not surprisingly, detailed weather data are not available for most of this large, sparsely populated area, but there are meteorological records that extend back for long periods for several places. The northernmost city in Chile is Arica; the average rainfall there between 1987 and 2002 was three millimetres per year. But that average is misleading because over one-third of the total rainfall in that sixteen-year period fell on a single day. It rained on a total of just fourteen days in those sixteen years, and in an earlier period, not a single drop of rain was recorded for fourteen years in a row. It seemed that the place where I was now stuck was drier than Arica, and I had only my broken-down truck to keep me company as the hours passed by.

This was not the first time I had got into a bit of a pickle while doing fieldwork. For my Ph.D. at the University of Toronto I had studied geographic variation in bee social behaviour, obtaining samples all the way from cold temperate Ontario to the subtropical climes of the Florida Keys. At one point I drove my car into the Okefenokee Swamp at dawn trying to get to the next sampling site in time.

Since then I have authored or co-authored over one hundred research articles on bees, most of them since becoming a professor of biology at York University in Toronto, Canada. Over the past thirty-five years I have travelled to all continents except Antarctica (where there are no bees) because of my fascination with these essential and beautiful little insects. Arid lands are my preferred destinations for the simple reason that bee diversity is higher in semi-deserts than in any other type of habitat. The normally dry and sunny weather is appealing to bees, which don't like to fly when it is raining or cloudy. Although my earliest research was mostly on bee behaviour, I have become increasingly aware of the need to promote the conservation of bees. Research performed in my laboratory demonstrates that bees are at a much higher risk of extinction than most other organisms. And that's a concern because the world as we know it would not exist without the pollination activities of bees: not only would there be few wildflowers, but our food supply would be substantially reduced. Some almost essential items—such as coffee—would be at risk (though many would consider coffee absolutely essential). So it is extremely important to increase our understanding of bees and to spread the word about these valuable creatures as widely as possible.

Consequently, I now put considerable effort into increasing general awareness of the significance of bees. I have, for example, written a pamphlet on the bees of Toronto; taught bee-identification courses in Ontario, Arizona and Kenya; written identification guides to the bees of Canada and adapted previously published ones for the needs of the Food and Agriculture Organization of the United Nations.

It is because of my interest in the conservation of bees that I was in the middle of the Atacama Desert with no shade but that underneath the truck, where I had spent quite some time. Fortunately, I had a large drum of water—essential for anyone travelling in this part of the world—and a drum of gasoline. I knew of other entomologists who had run out of gas and been stranded in this region for days before anyone else passed by. Without water and gas, you would not last long. I imagined the headline: "Mummified Bee Biologist Found in Desert."

Large amounts of water and gas are two things everyone needs in this beautiful wilderness. But few take a drum of liquid nitrogen with them as well. Why would I take liquid nitrogen into the desert? At -196°C, it was certainly not for cooling beer. The liquid nitrogen was for storing bees in a way that prevented their proteins from breaking down. Why was I interested in the proteins of desert bees? Why was I risking life and limb studying the proteins of insects so inconspicuous that even if you had thousands of them nesting in your lawn you would probably not know it? The answers to these questions form part of an extended narrative that eventually led me to conclude that bees may be the proverbial canaries in the coal mine of the globe's terrestrial habitats. Not only do I believe that bees can tell us much about the state of the natural world, but I also believe they are particularly good at indicating the state of the environment in areas that have been considerably influenced by human activity. But in case you think bees are unimportant, or even perhaps mostly a nuisance (after all, some of them sting), let's imagine what the world would be like without them.

Let's start with the obvious: there would be far fewer flowers. Plants have sex through the transfer of pollen from the male part of one to the female part of another. Pollen may be transported on the feathers of a hummingbird, the hairs on a fly or the tongue of a bat, and lots of it is transported by the wind (you may have an allergic response when plants are having sex using aerial currents as their intermediary). But the pollen of most flowering plant species is transported on the hairs on the body of a bee.

In the absence of bees, most flowering plants would not persist for very long. Yes, there would be some flowers left. The elongate red ones that are pollinated by hummingbirds; the large saguaro cactus flowers that are pollinated by bats; the white, moth-pollinated orchids that show up so well in the moonlight; and the dull flowers that give off a scent like rotting meat to attract pollinating flies—all of these flowering plant species could survive without bees. But the world would certainly be a less joyful place for us if the only flowers were on cactuses or smelled unpleasantly stinky.

In the complexity that is the web of life, we rarely understand the extent to which the continued existence of one species is dependent upon the presence of another. When it was a mere seedling, did that bat-pollinated cactus require the shade provided by a bee-pollinated flowering plant to avoid shrivelling up in the dry desert heat? If so, even the cactus might disappear from the face of the planet if bees were no longer around. We simply do not know enough about ecological interdependencies to understand what proportion of the organisms on the planet rely, directly or indirectly, upon the pollination activities of bees.

Certainly the loss of all bees would result in catastrophic cascades through the terrestrial ecosystems of the world. If many of the flowering plants were to disappear, the other species that rely upon those plants would also be in trouble. How many squirrels would there be without the nuts that result from pollination by bees? How many songbirds would there be without the berries that result from pollination by bees? No squirrels and no songbirds means no predators that eat the squirrels and songbirds. So the impact of bees extends throughout the food web—even to us.

We are part of the food chain (usually the end link, because comparatively few people get eaten). We are also a very large part of the global food web, appropriating perhaps one-quarter of the entire ecological productivity of the planet. Not all our use of the world's ecological productivity is through food. We cut down rainforests and drain peatlands to feed our insatiable demand for biofuels. We grow cotton for clothing, harvest wood for construction, and produce coffee to help us get going in the morning and sedatives to help us get to sleep at night. All these commodities rely to some extent upon the pollinating activities of bees.

"If the bee disappeared off the surface of the globe, man would have only four years of life left." That quotation has been attributed to Einstein, although scholars can find no evidence that the unkempt sage said anything of the sort. I think some, perhaps most, of us would survive for longer than four years without bees, but there's no question that the food supply would be substantially reduced. Why? What exactly is the impact of bees upon our food supply?

Consider breakfast. Eggs, maybe a slice of watermelon, toast with butter and jam, and a cup of coffee with a dash of milk—all are common components of a North American breakfast. The only item in the list that bees do not play a direct role in producing is your toast, since wheat is pollinated by wind.

Eggs come from chickens, and chickens eat seeds, among other things. Many of the seeds in the diet of a chicken would not be produced in the absence of pollinating activities of bees. And watermelons are entirely bee-dependent. Each fruit requires more than a thousand grains of pollen to be produced. (It's lucky for us that bees are such messy shoppers and leave behind so much of the food they collect!)

Coffee bushes do not need a pollinator to produce beans because they can self-pollinate. But when a bee moves pollen from one bush to another, the yield increase is enormous. Eggs, coffee and most fruits and vegetables would be a lot more expensive without bees because there would be much less of them.

What about the milk for your coffee and the butter for your toast? The cows that produce milk get most of their nourishment from grasses, which are wind-pollinated. But grass does not grow during the long winters at temperate latitudes, and alfalfa is one of the most important winter forage foods for cows. Alfalfa requires pollination to produce seed for next year's crop. A substantial proportion of our meat and dairy products would not be possible without the bees that pollinate alfalfa.

Last we turn to jam. Most jams come from berry crops, such as strawberries, raspberries and blueberries. These plants generally grow berries only after having been pollinated by bees.

Am I suggesting that we would all starve in an almost colourless and stinky world if it weren't for bees? No, that would be an exaggeration. Many of the world's staple crops are wind-pollinated, including various cereals, rice and corn. Others crops are pollinated by insects that are not bees; chocolate, for instance, is produced from cacao, which is pollinated by midges, and figs result from pollination by tiny wasps. I like chocolate and I like figs, but I wouldn't like to have to rely upon them for most of my non-starchy nutrition.

You may be thinking that we need not worry about any of this because our industrialized agriculture has an industrialized

pollinator to take care of all our pollination needs: the western
domesticated honey bee.[1] But recent developments suggest that
we cannot be so confident in this one major pollinator; hardly a
week goes by without the media mentioning the problems faced
by these bees and their keepers.

Concern about pollination and pollinators became so intense
that the National Research Council (NRC) in the United States
commissioned a report on the status of North American pollinators. The report, published in 2007, included a graph showing the
change in the number of managed honey bee hives in the United
States from 1945 to 2005. It's a scary graph. Apart from a few wob-
bles here and there, it shows a more or less continuous decline in
the number of hives. If this downward trend continues unabated,
there will be no honey bee colonies in the U.S. by somewhere
between the years 2040 and 2060. Of course, this is somewhat
absurd: the principle of supply and demand suggests that as the
number of hives decreases, each one will become more valuable,
and so the rewards for keeping them should increase. (But the
report was written before the latest calamity to impact honey bees
and their keepers, the mysterious colony collapse disorder, struck.)

Still, I suspect that things might be worse than even the
depressing scenario outlined in the NRC report suggests, as our
need for pollination is increasing. A recent survey of food production since 1961 shows that the number of pollinator-dependent

---

[1] Several of the approximately nine species of honey bees (the exact number
is subject to debate) have been domesticated, but only one is found almost
worldwide. As this species, properly called the western domesticated honey
bee, is the only one discussed at length in this book, I will use the shorter and
more commonly used term "honey bee" to refer to it alone.

foods has quadrupled in less than fifty years. The same study noted that this increase far outstripped the rate of growth of the world's domesticated honey bee hives, suggesting that either wild honey bees (escaped, feral colonies) or other pollinators have been helping us without our being aware of it. This growing demand for pollinator-dependent crops could cause a crisis in pollination.

The increased need for pollinators is particularly acute in almond orchards. Almonds are a needy crop in terms of pollination, and every year over half of the honey bee hives in the entire United States are taken to California to pollinate this one crop. In 2005 there weren't enough hives available, so some bees were imported from Australia. The area of land covered by the crop is expected to increase by 50 percent over 2005 levels in the next few years, so the pollination needs of the almond industry will only grow. This single crop will soon require more than two million honey bee hives each spring. That's over 70 percent of all the managed honey bee colonies in the United States. Having most of the domesticated honey bee colonies in the U.S. gathered together in just one state for the pollination of just one crop sounds like a recipe for disaster.

The question of why American beekeepers are in such trouble is complex because so many factors have impacted the industry. It's depressing to consider the many issues that have emerged in the past thirty years. First there were the problems with Africanized bees—originally called killer bees (that epithet was dropped because it was considered too scary). The Africanized-bee problem in North America had its genesis in the escape of a particular form of honey bee native to Africa. This strain was both aggressive and hard-working. Warwick Kerr, a Brazilian bee biologist, wished to crossbreed these bees with the domesticated variety that had been taken to South America hundreds of years earlier. The Brazilian bees were considered fairly peaceful but not very productive. Kerr hoped to transform the somewhat lazy Brazilian insects into harder-working bees through the introduction of genes from the African strain. He hoped that peaceful would beat out aggressive and that hard-working would beat out lazy in the genetic lottery of interbreeding. But in 1957, some Africanized bees escaped in southeast Brazil, a long way away from the almond growers of California. Living pests are not like pollutants, however, which can be turned off at source; once introduced, they reproduce and can increase unaided in both number and geographic range. Furthermore, these increases are usually almost impossible to stop. The escaped African strain gradually took over most of South America, then Central America, then Mexico, and by the late 1980s, it was on the verge of invading the United States.

These aggressive bees are more time-consuming to manage than the other commonly used honey bee strains, and that can make a marginally profitable industry economically inviable. Consequently, the United States Department of Agriculture spent vast amounts of money trying to come up with ways of preventing the arrival of Africanized bees in the U.S. The invasion was stalled for a while because habitats at various points along the narrow Central American isthmus were unsuitable for the advancing occupation. But the inevitable eventually occurred, and the first Africanized bees in the U.S. were detected in Texas in 1990. They are now in almost every state along the southern margin of the U.S., as well as Nevada, Oklahoma and Arkansas; they reached Florida in 2002. They do not get in the news much these days; their economic impact was originally worse than their sting, and beekeepers have since had worse problems to overcome.

While the North American honey bee industry was anticipating the arrival of aggressive bees, nasty bee parasites—specifically, tracheal and varroa mites, both accompanied by a bunch of diseases—appeared. Tracheal mites are tiny parasites that live in the breathing tubes of bees (imagine having cockroaches crawling around inside your lungs). The debilitating effects of these mites on honey bee colonies were first noticed on the Isle of Wight, off the coast of southern England; the mites spread from there to the rest of the British Isles and then to other parts of the world. They reached North America in 1984. Tracheal mites are difficult to detect because they live inside the bees. In fact, beekeepers often don't know that these pests are in their hives until the infestation is so severe that the colony becomes considerably weakened. Once inside a colony, the mites will stay until eradicated. Recent methods for control of the mites involve putting sugar with vegetable oil or shortening into the hive. Greasing the bees in this manner makes it difficult for the mites to transfer from one host to another. This interesting approach at pest control doesn't involve the use of unpleasant chemicals.

There are two species of varroa mites, only one of which, appropriately named the destructive varroa, is causing large-scale problems for the honey bee. These large parasites can be found on the surface of adult bees or on pupae and larvae, where they suck the bodily fluids of their hosts. For us, they would be like a mosquito the size of a small dog attached semi-permanently between our shoulder blades, where we can't reach it. Originating on a different honey bee species in Southeast Asia, where they were first noticed in 1904, these natural enemies of domesticated bees were detected in North America in 1987. They can have a debilitating effect on colonies, and it seems that the almost complete disappearance of wild (as opposed to managed) honey bees in much of the New World can be attributed to them. The hives can be treated with formic acid, a chemical related to vinegar, to get rid of these parasites, but for obvious reasons its use is not recommended during honey production.

The other enemies of our industrial-scale agricultural pollinator include small hive beetles, wax moths (which eat the brood comb), at least two bacterial diseases, two fungal diseases and no fewer than eight viral diseases—one of which has been implicated in the latest round of catastrophes to affect honey bees and the people who keep them: colony collapse disorder, or CCD.

The main symptom of CCD is the mass disappearance of bees during the winter. Beekeepers inspecting their hives in spring find them almost entirely empty. CCD caused the loss of over 40 percent of the colonies in the U.S. during the winter of 2006–07, with some beekeepers losing more than 80 percent of their hives. Losses continue, with 36 percent of hives dying out over the winter of 2008–09. CCD has been blamed on everything from pesticide use to the bees being disoriented by signals from cellphone towers; a recent headline read "Mobile Phones Are Killing Bees—Mankind Will Be Extinct in Four Years!" I favour this hypothesis for the simple reason that I do not like cellphones. But when pressed I will admit that cellphone towers were around for a long time before CCD was first detected, so the facts don't fit the hypothesis. Most recently, another bee disease, Israeli acute paralysis virus (IAPV), has been blamed for CCD. Its causative role has been demonstrated experimentally, but there are multiple strains of the disease and some seem to have little effect upon bees.

Another view is that CCD should be called MSD, or multiple stress disorder. Stresses on honey bees have increased for many reasons—not least because of the numerous treatments for the numerous ailments that now afflict them (being repeatedly doused with vegetable fat and vinegar would certainly make me feel less vigorous). A range of novel pesticides are now in widespread use in North America, despite having sparked lawsuits against pesticide companies by European beekeepers. Transnational transportation may also have had an effect. During their journey from crop to crop and state to state, the poor bees are exposed to multiple pesticides and other environmental contaminants, as well as the stress of travel. The bees are also given a series of different monoculture crops to feed on when they're being moved like this. How healthy would we be if in March we ate nothing but almonds, in April nothing but apples and in May nothing but blueberries? This is what we force many honey bees to do (albeit on the pollen rather than the fruit).

With this seemingly continual onslaught, we should have sympathy for our beekeepers. Not only have they had to cope with these nasty parasites and diseases, but they have also had to deal with very low prices for their product because the North American market has been flooded with cheap honey from overseas. (While being "flooded" with honey might sound nice, the quality of many imports is poor and they often have high concentrations of pesticide residues.) As a result, North American beekeepers now make most of their income from renting out hives for crop pollination rather than from selling honey.

Whatever the cause or causes of CCD, it is yet another indication that relying on just one organism to provide almost all our pollination needs is an unwise strategy. Just as we protect our finances by not putting all our funds into a single stock, we should hedge our pollination bets by not relying so heavily upon the activities of a single species.

There is another reason why concentrating our pollination efforts upon this one species is not a good idea. Please don't tell any beekeepers I said this, but honey bees are not always good pollinators. Your hypothetical breakfast is a case in point: honey bees are poor pollinators of alfalfa and some berries (blueberries, for example), and although they do a good job with watermelon, wild bees sometimes do the work well enough that honey bees aren't needed. Alfalfa flowers have a special mechanism that must be "tripped" by pollinators, and honey bees are not very good at getting this mechanism to work. Blueberries keep their pollen hidden inside minute anthers similar to salt shakers, and honey bees don't know how to shake the pollen out. Both of these crops are far better pollinated by wild bees than by honey bees. But as we shall see, all may not be well for these other bee species either.

It is these other bees that most interest me. There are over 19,500 described species; they are beautiful but largely ignored, and without their unseen activities, the world would be a far poorer place, aesthetically, ecologically, economically and nutritionally. It is not only the pollinating role of these unsung heroines (and it is the females that do almost all of the pollination) that makes them important. Wild bees are particularly sensitive indicators of the state of the environment. They are especially good at reflecting the state of environments that have been heavily modified by us; indeed, I believe they are better monitors of this than any other creatures on earth.

Environmentalists often point to declines in the numbers of birds and mammals as indicators of our negative impact upon the planet. Spotted owls tell us that we have reduced the Pacific coast rainforest to fragments that are too small; polar bears are beginning to tell us that we have caused too much arctic ice to melt; the giant panda can tell us that too much wild bamboo habitat has been turned into farmland. But large charismatic vertebrates like owls and bears (polar or panda) are the tiny tip of the iceberg when it comes to biological diversity. There are more species of bees than there are of birds and reptiles combined; there are approximately as many species of bees as there are of birds, mammals and amphibians put together; there are more species of bees and wasps combined than there are of plants. There is power in numbers. We can better estimate changes in ecological conditions with insects than we can with the more popular birds and mammals simply because there are so many more species to give us the information we need.

Most of the land cover of planet earth is already dominated by human activity through agriculture, silviculture and urbanization. We need to be able to discriminate between diverse, ecologically healthy habitats and the more severely stressed areas when both have already been considerably modified by our activities. This book is largely an attempt to outline why I believe wild bees are such superb organisms to use for these comparisons. We will have to take some diversions along the road, but fortunately, these diversions usually provide me with excuses to talk about bees, and I hope you will enjoy reading about them almost as much as I have enjoyed finding out about them.

2

The Future of Our Food

"Yes, I have a guitar, but it's not a very good one. Why do you ask?" John MacDonald was rightfully perplexed. I had hired him to conduct blueberry-pollination research, not to sing for his supper. But at this stage of our work, it was essential that we find a guitar right away.

John was a high school student and so was quite pleased to have found a summer job in Cape Breton, an economically depressed part of Canada where opportunities are few and far between, especially in the wilds where the highland blueberry company was located. Two friends of mine, Bill and Fenella Nicholson, had organized a group of local landowners into a collective of organic blueberry growers. They had asked me to find out whether the wild bees in their area were capable of doing a good enough job of pollinating the crop in the absence of rented honey bee colonies. So I set about the task and hired John as an assistant.

Pollination of a flower occurs when pollen is transferred from the anther (the male part) to the stigma (the female part). Some plants can self-pollinate within an individual flower. Self-pollination can also occur when pollen from one flower is transferred to the stigma of a different flower on the same plant, but this happens only with the help of an agent. Cross-pollination occurs when pollen is transferred between different individual plants. To determine what's needed for an economically viable crop, we must find out whether self-pollination is as agriculturally productive as cross-pollination.

Establishing whether a plant is capable of self-pollination within an individual flower is relatively straightforward—you put bags over the flowers before the buds open and leave them there until the petals fall off. If, later in the summer, there are no berries on the previously bagged parts of the plant, then there was no self-pollination within individual flowers. It's more complicated to discover whether pollen from one flower can pollinate other flowers on the same plant; this requires experimental collection and transfer of pollen. The added problem with blueberries is that collecting their pollen is not easy. Sexually, blueberry plants are coy; they hide their pollen.

In most plants, pollen forms on the outside of the anther, making it comparatively easy for the bee (or another pollinator, the wind or even a biologist's assistant) to get at it. If you have left lilies in a vase for a while, you will have noticed that their powdery pollen falls easily from the anthers, dusting the large lower petals or the table beneath. But blueberries are different; their anthers are hollow and the pollen is on the inside, where it has to be shaken out. This makes the experimental collection of blueberry pollen unusually complex: it involves a tuning fork, preferably one tuned to C.

Bees extract pollen from blueberry anthers by buzzing the flowers. They hold on with their legs or mandibles and vibrate their wing muscles rapidly—the bee equivalent of shivering. These vibrations shake the pollen out into the air, where it gets caught on the hairs of the bee's body. From there, the pollen grains meet one of two fates: either they get eaten by a bee or they stick on the stigma of a female plant when the bee visits another flower. How can a human being possibly collect this pollen? Holding onto the plant with your legs and waving your arms around isn't going to work, even if your legs are as hairy as mine. This is where the tuning fork comes in.

I had put John to work with a small plastic vial in one hand and the tuning fork in the other. He'd spent several days hitting the tuning fork against a rock, then holding it against the base of a flower with the vial positioned over the petals. This required some considerable coordination, but it was possible and it actually worked. Pollen shot out of the flowers and collected as a fine, pale yellow dust on the inside of the vial.

At times it had been difficult for poor John, a teenage lad playing with a tuning fork in a field of flowers. His friends thought he was going a little strange. But now things had got worse: just before the peak of blueberry flowering, the tuning fork broke. It would take quite some time to get another one—at least a day's return drive—and to compound the problem, we were now at the peak of flowering, and each day was crucial. Hence my query about the guitar.

"Yes," John had said in answer to my question. "Why do you ask?"

"You can pluck the string that gives the same note as the tuning fork, hold it against the flower and it will shake out the pollen just as well."

Now, much to the mirth of his friends, John had to sit in the blueberry fields playing his guitar to the flowers until another tuning fork arrived or the flowers stopped blooming.

In a world without bees, these are the lengths to which we'd have to go to get blueberries for our pancakes or our jam. Even the most musical of blueberry lovers would draw the line at playing the guitar to a field full of flowers for weeks on end to ensure a good crop. In fact, you would probably need all the guitarists in the world to replace the blueberry-pollination activities of bees, and that's just for the crop in Nova Scotia. Exactly this situation has arisen in parts of China, where pesticide use has wiped out bees in orchards and the flowers now have to be pollinated directly by human hands, aided by special pollinating paintbrushes.

Our melodious little research project demonstrated that pollination was essential for the blueberry crop, and that the wild bees could do the job as long as the weather was good enough during the flowering season. The proportion of flowers that resulted in fruit varied considerably from year to year. A cold, wet and windy June would result in poor yields; warm and sunny weather would result in a good crop. Unfortunately, the weather in this area was unpredictable, making it difficult to run the blueberry cooperative profitably.

More reliable yields are obtained on mainland Nova Scotia and in other slightly warmer locations in maritime Canada and the northeastern United States, where blueberries are often adequately pollinated by wild bees. But these native pollinators can do an adequate job of pollinating the crop in these areas only if the fields are small. In a large field, the bees will do a fine job of pollination around the periphery, but comparatively few will venture into the middle. Although some will nest near the centre of the field and pollinate the blueberry flowers there, they won't reproduce very well because as soon as the flowering period is over, they'll have to commute to other flower species at the edge of the field for food. The problem here is that a large blueberry monoculture provides a feast for bees when the flowers are in bloom but a famine when they are not. An additional complexity is that blueberry management involves burning or otherwise disturbing the ground every two years. The result is that there are almost no flowers in alternate years, and without other nearby food sources, bees will find themselves with a feast one year and a famine the next.

Bumble bees are common visitors to blueberry flowers. As social bees, they are active whenever the weather is warm enough, so they need pollen and nectar from spring until fall. They can forage from blueberry flowers for perhaps two or three weeks, but they need additional resources for the months when blueberries are not in bloom. Bumble and other bees that need food when blueberries are not flowering have to either live around the edge of the field, closer to their alternative food sources, or commute long distances. For wild bees, expending energy on commuting means there is less energy available for other tasks, such as constructing a nest, laying eggs, collecting food for offspring, overall personal maintenance and performing house repairs. For us, a longer commute and higher gas prices are inconveniences, but they don't generally result in our having fewer offspring, thereby causing our population to decline. For the bees, the consequences of a long commute are more dramatic. If the energy expended on commuting to food sources becomes too great, the bees cannot collect enough food for their offspring and their population will decline. It may decline to zero.

~

Alfalfa is used as food for a wide range of livestock. Most important, the hay made from it is used as winter fodder for cattle. Some alfalfa is also used as seed for turkeys and chickens (so your eggs may have benefited from pollination of alfalfa flowers by bees). You may also enjoy alfalfa sprouts in a salad or as a juice.

Alfalfa is native to the Near East but was introduced into Europe (where it is known as lucerne) so long ago that we cannot be certain of the timing. Early settlers of North America brought it to the shores of New England in 1736, but the climate there was a bit damp for it. Greater success was achieved on the prairies, where alfalfa was introduced (from Mexico, where Spanish conquistadors had taken it as fodder for their horses) in the 1830s.

After initial success, alfalfa production on the prairies declined. Farmers planted such huge areas of the crop that they destroyed much of the nesting substrate needed by the native pollinators. Further problems arose with the use of pesticides in the 1950s, and for a while alfalfa pollination was so poor that Canada had to import alfalfa seed, even though a decade earlier it had been an exporter. Alfalfa production in North America was largely saved by the alfalfa leafcutter bee.

These oblong-shaped, greyish-brown bees got their name because of their preference for alfalfa blossoms, which have a complex tripping mechanism that they can handle easily. As it searches for nectar, an alfalfa leafcutter bee exerts pressure on the lower petal, where a spring-like mechanism causes the anthers of the flower to brush against the bee. When the next flower to be visited is tripped, its stigma contacts the bee and pollen is transferred. The alfalfa leafcutter bee is particularly good at doing this. Many other species of bees visit the flowers, but most, including the honey bee, are not capable of detonating the mechanism correctly, so pollination does not occur, or at least does not occur often. In Alberta, honey bees pollinate fewer than one in one thousand of the alfalfa flowers they visit, whereas alfalfa leafcutter bees have a success rate of almost 100 percent. A field with managed populations of alfalfa leafcutter bees will produce an average of 350 kilograms of alfalfa seed per hectare, seven times as much as is possible without them. Some producers can achieve yields up to three times greater than that. Current prices for alfalfa seed can reach three dollars per kilogram. Based on that price, the value of the increased yield resulting from these bees is about nine hundred dollars per hectare.

But the economic value of alfalfa lies not only in its seeds—alfalfa hay is also enormously important as winter feed for cattle. Prices for alfalfa hay were fairly constant earlier in this decade, averaging just under $100 per ton. But in recent years prices have almost doubled, and in August 2008 they reached a record high of over $180 a ton. If we include the added production of dairy, eggs and meat in the value of alfalfa, its annual worth in the United States in 1999 was $145 billion. It is probably twice that amount now. In 1983, it was estimated that alfalfa was responsible for 10 percent of beef production in the U.S. So without alfalfa pollinators, there would be much less meat to eat. Even purely carnivorous people owe a lot to these little unsung heroes of pollination. But how are they managed?

With the destruction of the native prairie, alfalfa growers soon realized that a lack of pollination was a problem, and as early as the Second World War, W. D. Clarke of White Fox, Saskatchewan, attempted to increase the populations of alfalfa leafcutter bees by drilling a hundred holes in white spruce logs. Most of the logs made up the walls of his cabin, so he was clearly concerned about the survival of this important pollinator. (I would encourage all of you to help tunnel-nesting bees by drilling some holes into the wood or brick of your own homes.)

The alfalfa leafcutter bee used to be found only in Eurasia, but it's now found throughout almost the entire northern hemisphere. It is also common in parts of Argentina, Chile, Australia and New Zealand. Many countries can trace their domesticated stocks of the species back to Canada, where domestication began in earnest in the 1960s, twenty years after Clarke first drilled those holes in his cabin walls. The bees continue to flourish today. I find them in my own backyard in Toronto, where they forage on many different flowers. Current alfalfa leafcutter bee management is a multimillion-dollar industry in many parts of the world. Domestication methods developed for the bee in North America have even been exported back to its native range, as have North American–raised bees, in an attempt to increase pollination, and therefore yields, of alfalfa crops in areas where bee and plant have coexisted for centuries.

Alfalfa leafcutter bees nest in pre-existing holes and line their brood cells with cut pieces of leaf (hence the name). They make their brood cells in a row; the entire nest's contents look like a cigar. These bees can be tempted to nest inside holes drilled in wooden blocks, blocks of polystyrene foam, the log walls of a cabin, bamboo poles and poorly repointed brickwork. Like many people in cities the world over, these bees are happy enough living in apartment blocks, and they are easy to manage in similar structures. In this case, the apartment buildings are cupboard-size conglomerations of foam blocks, each with dozens of nesting holes drilled into them. A roof over the entire structure keeps away the worst that the elements can bring. Thousands of female alfalfa leafcutter bees can nest in these crowded conditions.

When we live in an apartment building, we locate our individual residence through the floor and suite numbers. It would be difficult to do this if we weren't numerate. Because bees can't count the way we do, they don't find their own nests by counting the number of rows up from the bottom floor. Their acute sense of smell is doubtless important in permitting them to find their way home, but if there are thousands of different nest odours in a small area, even that may not help. To make it easier, beekeepers paint complex patterns on the surface of the hive blocks so that a returning female can tell, from some distance, where her nest is. Some blocks have the bee's scientific name, *Megachile rotundata,* stencilled across the front in capital letters. I doubt the bees are able to read Latin, of course, but they can discriminate patterns well enough to be able to tell where their nest's entrance is among the letters of their scientific name.

The Canadian prairies provide excellent habitat for alfalfa leaf-cutter bees. There is a ready supply of alfalfa, and the summers are perfect for them—although the winters do pose a challenge. To keep them from freezing to death, the beekeepers store them in coolers that are at a temperature warmer than ambient (between 3°C and 7°C; your kitchen refrigerator would be fine for this purpose). The cigar-like rows of brood cells are extracted from the nesting blocks towards the end of summer for transfer to the cool room. At this time, the bees are fully fed larvae hibernating inside their cocoons. The cold-room temperatures are warm enough to prevent the bees from freezing to death but cool enough to keep them in a dormant state. If they are kept too warm during the winter, they may emerge too early, before the alfalfa is in bloom.

This pattern of overwintering makes it possible for the keepers of alfalfa leafcutter bees to time the emergence of the adults to coincide perfectly with the development of the alfalfa flowers. They do this by moving the larvae from the refrigerator to an incubator. If the incubator is kept at 30°C, the bees will emerge three weeks later. As long as the farmers can predict when the alfalfa will bloom, which is moderately easy with information from meteorologists, the beekeepers can synchronize the bees' emergence with the needs of the crop.

This bee's winter dormancy also makes it possible to ship cocoons to regions where the activities of alfalfa leafcutter bees are needed. Canada has exported these bees to places as far and wide as Romania and Mongolia. The cost is an absolute bargain at half a cent per bee. Alfalfa farmers can make as much money from the sale of excess bees as they do from the alfalfa itself. That's like having your alfalfa-seed cake and eating it too.

So that's how bees help produce our meat and dairy, but how do they provide our slice of watermelon? These enormous, delicious, extremely juicy fruits result when a single flower is pollinated by bees. In fact, to produce a large, healthy, beautifully oval watermelon, the flower needs to receive at least one thousand pollen grains. (With fewer grains than that, the fruit will be too small or misshapen.) Given their high demand for pollen, watermelons are a good subject for those studying the impacts of pollinators and growing systems upon fruit yield. This topic has been studied in detail by the biologist Claire Kremen and her colleagues.

Claire has a more diverse set of research experiences than most young academics: in addition to conducting groundbreaking work on pollinator biodiversity, she is also well known for her studies of the effectiveness of the national park system in Madagascar for conserving that nation's biological treasures. Her watermelon research was performed in California while she was based at Stanford University.

It is becoming increasingly important for us to be able to assess the impact of wild bees on agricultural production in both organic and traditional farms. Claire and her colleagues set the standard for such work, and they set it rather high. They studied watermelon pollination in three different groups of farms: organic farms that were close to reasonably large areas of natural habitat, organic farms that were far away from natural habitat and conventional (pesticide-using) farms that were also far away from natural habitat.

First, the researchers had to calculate how many pollen grains individual bees deposited during an average visit to a watermelon flower. This required painstaking observation of virgin flowers until they were visited by a bee. The observer had to identify the species of bee (something that we shall soon see is far from easy), carefully remove the flower and then count the number of pollen grains on its stigma. To get accurate estimates, the researchers had to repeat this process many times for each bee species. Next, they counted how many visits each bee species made to the flowers. This involved walking among the watermelon plants and counting the number of individual bees visiting the flowers. They had to do this for hours and hours, day after day, in the hot California sunshine. Only once all of these data were at hand could the researchers perform their analyses.

They found that no fewer than thirty-nine wild bee species visited the watermelon flowers at their study sites. (Add the honey bee for a round total of forty.) The impact of farm type upon pollen deposition was enormous. Watermelon flowers on organic farms that were close to natural habitat usually received enough pollen from wild bees to produce an economically viable crop, with an average of 1,750 pollen grains per flower; 80 percent of the farms obtained ample pollination from wild bees alone. In contrast, organic farms that were far from natural habitat did not always receive the required pollination levels, averaging only 700 pollen grains per flower. Only half of these farms could produce an economically viable crop with wild bees alone. But the situation for wild bees on conventional farms was even worse: only 250 pollen grains per flower were deposited by wild bees, and not a single conventional farm obtained enough pollination from wild bees for the crop to be economically viable.

These results were at least partly due to differences in the diversity and number of bees among the three farm types. The organic farms close to natural habitat averaged more than eight different bee species, and the wild bees visited each flower more than seventy times per day. The organic farms far from natural habitat had half as many wild bee species and fewer than thirty-five visits per flower per day, while the conventional farms had a similar number of species but fewer than thirty visits per flower per day.

Those calculations were for wild bees only; when the activities of the honey bee were included, all the farms received economically viable levels of pollination. Honey bees alone were sufficient for the crop on the conventional farms. But on the organic farms, whether they were close to or far away from natural habitat, honey bees were not up to the task, and wild bees were essential.

Another important finding emerged when the researchers compared data from two different years of study. In one of the years, all the bee species that visited the watermelon flowers were required for the crop to receive adequate pollination. In the other year, just two of the wild bee species were enough for economic productivity. This indicated that reliance on just a few wild bees would not be a good strategy in the long run. While only two or three species would be needed in some years, the entire suite of wild bees in the neighbourhood would be required in others.

Wouldn't it be useful if the wild bees that were best at pollinating the watermelon flowers were also the ones most resistant to the environmental changes taking place where this crop is grown?

Unfortunately, the opposite seemed to be the case: the bees that were most sensitive to habitat change were usually among the best pollinators. In California's watermelon fields, the first wild bees to disappear as the habitat decreases in quality are usually the most important ones for the crop.

Claire's study provided a cautionary tale. If another disease like CCD were to sweep through our domesticated honey bee colonies and annihilate them, only organic farms close to large areas of natural habitat would be able to produce a viable watermelon crop. This is more proof that reliance on one industrial-scale operator, the honey bee, for the pollination of all our fruit and vegetables is very risky.

~

Coffee is the most economically valuable legal crop. Indeed, it is the second most valuable legally traded commodity after oil. Although it would be rash of me to suggest that coffee is as important for the survival of Western society as oil, it does seem that much of civilization as we know it would grind to a halt without it. Given that coffee flowers are capable of self-pollination, it seems odd to argue that bees are responsible for the crop. But without bees, coffee production would be a fraction of what it is, and the price of coffee would be much higher.

Using methods similar to those described for blueberries (but without the musical accompaniment), it is possible to get yield estimates for coffee plants that are not visited by bees and then compare those estimates to the yields of plants that are. Results vary considerably for a wide range of reasons: there are three different species of coffee; plants are grown under conditions of varying suitability for bees; and different species of bees are found in different coffee-growing areas, and not all are suitable pollinators. This list of variables is far from exhaustive, so it's not surprising that the increase in coffee production because of bees can be anywhere from 15 to 50 percent. The loss of pollinating bees would result in perhaps a 33 percent reduction in coffee production.

The conservation biologist Taylor Ricketts authored the classic paper on the value of wild habitats that support coffee pollinators. At that time Taylor was at Stanford University, though he now works for the World Wildlife Fund. His website states that he "is an affable, fun-loving fellow who keeps a drum set in his office." (I'm glad my office isn't near his.) For his paper, he studied the economics of pollination and natural land conservation on one coffee farm in Costa Rica. This particular farm was surrounded by a complex assortment of mixed-use farmland and fragments of woodland. Some of the woodland fragments were quite large, of a size that had been shown in other research to harbour good populations of wild bees. Taylor wanted to know how important maintaining the remaining woodland was for the economics of coffee production.

Taylor studied coffee yields in parts of the farm that were adjacent to large forest patches, parts that were as far away as possible from these forests and parts that were somewhere in between. At each of the three groups of sites, he measured the yield of coffee beans on bushes that had been pollinated naturally and compared the results with those of bushes that had been pollinated by hand.

Taylor determined that the coffee flowers near the forest were pollinated perfectly by wild bees. For both naturally pollinated and hand-pollinated plants, the proportion of flowers that produced beans and the average weight of the beans were similar. But the situation was different far away from the forest patches. Here, hand-pollinated flowers produced 11.5 percent more beans and the average bean weight increased by 8.3 percent. Combining these figures, Taylor calculated that in the parts of the farm too far away for the forest-dwelling bees to reach, hand-pollination would increase yields by almost 21 percent. Not surprisingly, the coffee plants at the middle distance gave intermediate results. This meant that in the parts of the farm farthest from the forest patches, the coffee bushes were producing 20 percent less coffee than they would have if they were pollinated adequately.

We can look at these estimates the other way round. What would happen if the patches of forest adjacent to the coffee farm were developed for cattle ranching or turned into additional coffee plantations? This would leave all the coffee bushes on the farm far away from natural habitat, and at best, all the plants would have the same reduced yield. The complete destruction of the forest patches around the coffee farm would result in a loss of US$62,000 per year. That's almost $400 per acre of forest. Interestingly, the Costa Rican government supports landowners who leave forests on their land, giving them $42 per acre each year—a little more than one-tenth of the value the forests gave to the coffee farm in Taylor's study. But those patches of forest were likely even more valuable than estimated, because they were also adjacent to other farms that each grew a range of crops, including coffee. The bees from the forests would be adding value to those farms as well. This wild land is economically valuable, but the government pays far less towards its conservation than it is worth. Still, Costa Rica provides more support for farmers who keep part of their land wild than do the governments of most countries, which offer no economic incentives for conservation at all.

These calculations were made at a time when coffee prices were depressed. Current prices are approximately double those that Taylor used for his analyses. In all likelihood, the value bees and forests offer to the coffee crops will also have doubled. Despite this, the farm that Taylor studied now grows pineapples.

~

We've seen the impact bees have on some common crops. But what proportion of our food relies on bees? Not as much as we might think. Perhaps one-quarter of all our food results from pollination (one-third is a commonly quoted figure, but this is likely a slight exaggeration). The staple crops that provide the vast bulk of what we eat, items such as rice and wheat, are pollinated by the wind. We would get our calories if there were no bees, but perhaps not the vitamins and other nutrients that we need in smaller quantities. We would not be as healthy without bees, even if we could eat enough food to keep our stomachs relatively full. Medical nutritionists have long known that dietary diversity is important for health.

Some other crops, such as carrots and broccoli, do not need pollination after planting, but the seeds from which the plants are grown would not be available without bees. In addition, crops that can be propagated without pollination suffer reduced genetic variability. Agave, from which tequila is made, has suffered from the increased incidence of diseases as a result of lack of genetic variability in the managed crop. Adding sex to agave propagation increases the plant's resistance to disease. The price of tequila would be even higher in the absence of pollinators, even though propagation of the agave plants does not depend upon pollination.

Of course, many of our foodstuffs do rely upon the pollination activities of wild bees. We need to know what these bees are so we can ensure that their helpful activities can continue in the face of environmental changes. But even the scientists who performed these classic studies often had to use informal descriptions of the species they saw visiting the crops. The phrase "small striped bee" narrows it down to a few hundred species in Costa Rica and a few thousand worldwide. Does "big black butt bee" mean a large bee with an average black behind or a medium bee with a disproportionately large black butt? Even decades of practice may be insufficient for biologists to be able to identify a bee species on sight. But the problems are worse than this. Most people—and even, as we'll soon see, most biologists—don't really know what bees are.

3

Honey, Queens, Hard-working Workers and Stings: Misconceptions about Bees

STUNG BY BEE KILLERS ON THE ISLE OF WIGHT, UNITED KINGDOM

What excitement! I had caught a bee wolf, a wasp that collects bees as food for its offspring. I was excited because this was the first time this species had been found on the Isle of Wight, off the southern coast of Britain, in decades. Just a few hours earlier, while we were on the ferry across to the island from Portsmouth, I had asked my fellow surveyor, George Else, the bee expert at the Natural History Museum in London, whether he thought we might catch this species. "No," he'd replied unenthusiastically. "It hasn't been seen here for years." Even though bee and wasp hunters had visited many times in recent years, I'd caught the first one spotted in this part of the world in a long time.

I was also excited because the first specimen I found had stung me. Most people would not be "excited" by being stung, but for me this was a fascinating intellectual experience. I had been stung by the ferocious bee-killing wasp!

I nonchalantly made my way over to George, at the other end

of the sandy cliff, trying to act cool. It just isn't "British" to show enthusiasm (or pain) to the nation's expert.

"Caught any bee wolves?" I asked calmly.

"Yes, actually," George replied. "Three."

Darn, I thought. "Have you been stung by them?"

"Stung?" he replied incredulously. "Why, of course not." Now I knew I was one step ahead of George.

"I have, and it was distinctly unimpressive."

"Really?" George replied, feigning disinterest.

This story illustrates several interesting phenomena, among them the odd competition that takes place among entomologists when they are comparing their prowess in the field. A little bit of luck is all that may be needed for one individual to catch the rarest species of the day, but others in the group will re-evaluate their views of that person's abilities as a field melittologist.

The bee wolf is a wasp that catches honey bees for a living. It paralyzes the bees with its sting. This is what the sting was originally for—not to inflict pain on humans, but to inflict immobility on some poor insect so the wasp's larvae could eat living food. Stinging wasps paralyze animals so they stay fresh while they're being eaten by their offspring. They keep their food in a state of suspended anima-tion until the larvae have finished eating it. Imagine having a small flock of chickens in the kitchen, comatose but alive. Every time you are hungry and want to barbecue one, you simply go and get it. Rather macabre, but this is analogous to what the wasp ancestors of bees have been doing for over a hundred million years (albeit to caterpillars, flies, beetles or spiders rather than chickens or cows). Chemical paralysis of food is an energy-saving way of keeping it fresh. It also allows your defenceless and fragile youngster to feed on something living without risk of harm. After all, it's better to have your chickens stunned in the larder than flapping around in your toddler's bedroom. It is possible, though, that some poor sting recipients may actually retain the sensory capacity to feel the nibbling of the wasp larva as they are eaten. Imagine being able to feel almost every last bite from the grub that is eating you. (Don't worry, vegetarians; the story will get better soon.)

~

When thinking about bees, most people conjure up a number of things in addition to stings: honey, complex social lives, hard work. The order with which these items come to mind may depend on whether you have a sweet tooth or have been stung recently. But in truth, this combination of features is found in far less than one percent of the species.

If you have been stung by a honey bee, you would likely be surprised to hear that they rate in the middle of the sting-pain index. Several scientists have spent considerable lengths of time, and experienced considerable pain, investigating how potent the stings of different insects are. Imagine being so fascinated by this subject that when you find a ferocious insect whose venom you have not yet experienced, you immediately grab it and get it to sting you. This is a dedication to science that is almost unsurpassed.

The desert field biologist Justin Schmidt started inflicting pain on himself while he was a graduate student in Georgia. (This was highly unusual, as it is normally the supervisor's job to inflict pain upon the students.) Justin looks like what you might expect a desert field biologist to look like: lean and wiry, with a complexion suggestive of long periods spent outside in dry heat. Even his voice sounds somewhat cooked. But perhaps the deep and cracked voice is a result of overuse of his vocal cords in research-induced screaming: he has invented an index of the potency of different species' sting venoms. Honey bee stings, according to Justin, come near the middle of the pack. If the feeling of a honey bee sting is similar to being gently prodded by someone's fingertip, then it would take a punch with a closed fist to approximate the excruciating pain that is inflicted by a bullet ant. Justin describes that sting as "pure, intense, brilliant pain. Like fire-walking over flaming charcoal with a three-inch rusty nail in your heel." The pain lasts for twenty-four hours.

Despite its middling ranking on the pain scale, a honey bee sting is something you likely will not forget. The stinger of a honey bee worker has large barbs on it, like a fish hook; the bee uses these to anchor the stinger in flesh. The entire sting apparatus is pulled out of the body of the bee as it flies away. Justin describes the pain as "like a match head that flips off and burns on your skin," though my experience suggests that this is a bit of an understatement. If you can remain calm during this mid-level pain experience, you may be able to watch the venom sac contract as the pain-inflicting fluid is pushed through the sting shaft into you. It may be painful to you, but the poor bee suffers death by disembowelment doing this, so have some sympathy for her.

Her? Yes, any bee that stings has to be a female. Males cannot sting, as the stinging equipment is a modified egg-laying apparatus. Males do not lay eggs, and therefore, by definition, they cannot sting. To understand the development of this sexual dimorphism in pain infliction, we have to go way back in time, before the origin of bees and stinging wasps, to the ancestor of the aculeate Hymenoptera.

Hymenoptera means "membrane-winged," and is a reference to the normally transparent membrane of the wings of bees and wasps and their relatives (as opposed to the leathery wings of grasshoppers, the scaly wings of butterflies and the crunchy forewings of beetles). Aculeate is the name biologists give to stinging membrane-winged insects (*aculeus* is Latin for "sting"). This ability to sting separates bees, wasps, ants and some of their more obscure relatives from the parasitic wasps and sawflies.

Two hundred million years ago, parasitic wasps used an ovipositor to place their eggs inside other organisms, such as caterpillars. These dastardly beasts had evolved from sawflies, which used their ovipositor (the structure of which includes saw-like blades) to slice into a leaf or a stem and deposit an egg. The ovipositor of the ancestral sawfly and its parasitic descendants has a long tube along which the egg passes. Glands that open into the base of the ovipositor produce lubricating fluids that ease the passage of the egg. But the ovipositor switched function from egg-laying to stinging somewhere in the early evolution of this ancestor of bees and stinging wasps and ants. The egg now comes out through a gap at the base of the sting instead, which freed the glands to produce paralysis-inducing compounds and pain-causing concoctions rather than mere lubricant.

The sting did not evolve as an offensive mechanism to inflict pain; it also did not evolve as a defensive mechanism. As we've seen, the original purpose of stinging was to paralyze another insect so it stayed fresh while being eaten. But pollen grains are defenceless and do not need to be subdued, so bees were freed from the need to paralyze their prey, and their sting came to serve a purely defensive function (this is true for the females of most, but not all, species of bees).

The waspy ancestors of bees had to be very quick and accurate with their stinging. This explains why more parts are involved in the orientation of the dagger-like shaft than are actually responsible for pushing in the venom. Consider the tarantula hawk as an example. These enormous spider-hunting wasps have a fearsome adversary in tarantula spiders, which themselves eat insects, including, if they get the chance, the tarantula hawk that stalks them. The tarantula hawk has to ensure that it can paralyze the tarantula before getting caught in its poisonous fangs. It does this by inserting venom into, or next to, the ganglia (nerve centres) that are between the bases of the tarantula's legs on the ventral side of its body. Unlike my two cats, tarantulas don't just flop down and turn over to get their bellies tickled. They fight to the death. The tarantula hawk has to strike like lightning so its venom can incapacitate its foe before the spider's gnashing jaws end the wasp's motherly devotion to duty. Once vanquished, the tarantula suffers the death of a thousand bites at the jaws of the tarantula hawk's grub-like son or daughter. This may take a week. The pain inflicted on people by these wasps is considered equal to that of the bullet ant. Justin Schmidt describes their stings as "blinding, fierce, shockingly electric. A running hair drier has been dropped into your bubble bath." Don't try this at home.[2]

When a honey bee worker loses its sting apparatus and venom sac, a greater volume of venom is injected into you. This makes the experience unusually painful, and the pain lasts a long time. The stings of most bees are comparatively weak and short-lived in comparison.

Of course, some bees have a strong sting, and I can feel the shaft going in even before any venom is extruded through it; the venom feels like fire. The pain may be detectable for half an hour or so, but it quickly diminishes in intensity. The bee's purpose is

---

[2] In fact, because some people are allergic to bee or wasp stings, it's wise to avoid getting stung by any of these insects.

not to inflict lasting pain but to ensure that it is released so it can live to reproduce another day.

But here is an additional complexity to the misconception about bees as stinging insects: there are many female bees that cannot sting at all. When a bee stings, the venom is pushed through the sting shaft by little valves that are attached to two narrow lancets. The whole structure works like a syringe, with the valves and lancets acting as the plunger. Many bees that cannot sting have lancets that lack valves or do not insert into the sting shaft. Even if these bees produced venom, they could not inject it. In some bees, the sting shaft has been reduced to a soft and blunt triangle that couldn't pierce the soft skin of a newborn baby. There is even a group called stingless honey bees (or just stingless bees, for brevity). These have minute vestiges of the sting apparatus, but they won't extrude them for you no matter how roughly you treat them. Their stingers are completely non-functional. They sound really nice. What could be sweeter than a stingless honey bee? Well, lots of things, actually. If you disturb them, dozens to thousands of these small bees will swarm around your ears, eyes and nose, biting sensitive membranes and vibrating their wings to increase the irritation. Some stingless bees will drool over you, and their saliva is a caustic mixture that burns the skin. But none of them can use its stinger to inject venom into you.

Some bees that have an intact sting apparatus, with functional lancets, valves and shaft, seem very reluctant to use them. Still others have stingers that are functional but too short to penetrate human skin. They may be able to sting you in places where your skin is thin and delicate, such as in your mouth or on a recent wound, but they cannot get through the skin on your fingers. They can probably sting the tongue of a potential predator, such as a bird or a small mammal, but you and I can pick these bees up without fear of harm. (It takes many years of practice to know which bees these are, however, so please do not try this at home, even if you know you are not allergic to stings.)

Of the more than nineteen thousand known species of bees, only the females of less than 75 percent of them are capable of stinging us, and most of them have a sting that causes pain for only a few minutes. This is certainly enough to ensure that you will let the bee go (unless you are a melittologist, as students of wild bee biology are called, or a masochist) and learn to avoid grabbing one in the future.

～

Honey is concentrated nectar. Bees collect nectar as an energy source, both for themselves and for the offspring. Honey bees have such complex and well-organized societies that they can plan ahead for rainy days (or many cold winter months), and they store large amounts of honey for such eventualities. Sorry to put you off your favourite tea sweetener, but honey is really bee vomit, in the sense that it's been stored in the bee's stomach for a while, at least while being transported back to the nest. But the view that all bees make honey is another misconception. In fact, less than 5 percent of them make any honey at all, and only a small fraction of those make honey in large enough quantities for us to be able to use it.

The bees that make honey are, of course, the honey bees (approximately nine species) and the stingless bees (fewer than five hundred species). The stingless bees do not put honey in combs, but instead store it in pots that look like bunches of grapes. Extracting their honey is either time-consuming (as it generally has to be sucked out of each individual pot using a squeegee-like apparatus) or somewhat destructive (as it requires cutting the honey pots and pouring the honey out from numerous shattered vessels). Stingless bees are tropical species found in all suitably warm regions. Because there are numerous species (more than four hundred), their honey is highly variable in flavour. But with a few exceptions (which we will discuss in Chapter 6), it is always more tasty than standard honey bee honey. I recently gave small samples of stingless bee honey to over a dozen biologists at a conference banquet; they all asked where they could get some more (alas, it is not generally available). Stingless bee honey also has medicinal properties, largely as a result of anti-bacterial action. Because of its flavour and health benefits, this honey is considered an exotic crop with mysterious properties for which new age tourists will pay a premium—if they can find some to purchase.

In most tropical regions, people keep stingless bee nests in their backyards to provide honey for just the one family. The Yucatan Peninsula is an exception. Here, ancient Mayan methods are used in a small industry that produces stingless bee honey. During the time of the Mayan empire, consumption of this honey was restricted to the upper classes because the delicious liquid was available in such small quantities. The bees were held in such high regard that the Mayans worshipped them. In one of the great tragedies of the last millennium, zealous, missionary-minded Western colonizers destroyed as much of the Mayan culture as they could. This reli-gion-based vandalism extended to almost the entire Mayan litera-ture: only three codices—books that are folded like a concertina—were saved from the inquisitorial flames. In parts of these codices, the bee god explains how to keep stingless bees, as well as how to extract the honey and propagate the hives. But with no surviving written literature, the Mayans had to pass down their beekeeping methodology orally over the centuries. Stingless bee culturing now rests in the hands of the few keepers of the Xunan Cab (as these bees are called in Mayan). The number of colonies being maintained in the Yucatan has decreased by 90 percent in the past thirty years.

Like stingless bees, bumble bees also keep honey in small grape-like pots, but the amounts made are so small you would be lucky to put a spoonful of it in your tea even at the height of colony development. In terms of bees making honey, that's it. Of the more than nineteen thousand species of bees, perhaps a dozen are honey bees, fewer than five hundred are stingless bees and fewer than three hundred are bumble bees. Only about 5 percent of the world's bee species make any honey.

What about the idea that bees are social insects? Bees are renowned for their complex social organizations, and hundreds to many thousands of individual bees can be found within a single nest. So it may come as a surprise to you that most bees are solitary. A solitary bee female constructs a nest alone, and her social interactions are restricted to mating with a male (usually only once) and laying eggs. This is not exactly the high degree of social complexity you might have expected. Only honey bees and stingless bees have the complex societies we typically associate with bees. Their colonies are perennial, with the queens living for several years. These societies have a very finely tuned division of labour, even among different groups of workers within the nest. The queen is never left alone; she never has to fend for herself, collect her own food, or help construct the nest (just as you won't see Queen Elizabeth shopping at the local supermarket or repointing the bricks at Buckingham Palace). All the food for the queen and her offspring is collected by the workers. As a result, queens and workers of "advanced" social bee species are readily distinguishable in both size (queens are larger) and structure (queens do not have pollen baskets).

Bumble bees, by contrast, live in annual colonies that last for only one spring, summer and fall. Only the potential queen survives the winter; she starts a colony in spring and initially produces just a handful of workers, but the colony soon grows to perhaps a few hundred. Later in the summer, the colony stops producing more workers and begins producing males and next year's potential queens.

Similarly short-lived annual colonies are formed by many species in an obscure group of little bees known as the sweat bees, so-called because of their propensity to land on exposed, sweaty human skin to lap up the nutrients. (Some people in the tropics call stingless bees sweat bees because they are similarly attracted to our cooling secretions.) Many people will swat them and get stung but believe that a mosquito has bitten them. According to Justin Schmidt's pain index, these stings are "light, ephemeral, almost fruity [sounds like wine] . . . A tiny spark has singed a single hair on your arm."

In total, perhaps one thousand species of bees have annual social lifestyles, with queens flying around in spring, workers in summer, and males and next year's queens coming out in fall. The complex societies that are found among honey bees, and that many people think of as typical of all bees, actually occur in fewer than five hundred bee species. Most bees aren't social but instead lead entirely solitary lives.

This brings us to two additional misconceptions: that bees have complex communication skills, and that they always live in hives. In fact, not even all species of honey bees live inside hives. Our common honey bee was easily domesticated because it lives inside hollow structures that we can mimic with artificial hives. But if we look at the dwarf and giant honey bees of Asia, we find that they build their nests on the outside of structures. I remember this well because I nearly ran into a colony of giant honey bees while looking for other bees in an orchard in India. These bees station layers of workers on the outside of the nest. Anyone foolish enough to approach too close to the colony will find hundreds of large, stinging insects ready to be eviscerated in a suicidal attack.

Still, the fact that these relatives of the domesticated honey bee live outdoors, rather than inside a hollow log, may help us understand the origin of one of the most marvellous feats of animal communication—the waggle dance. Returning honey bee foragers dance excitedly when they find a good food source. While dancing, the bees trace out a shape that looks like a hamburger with a squiggly line of mayonnaise down the middle. It's the pattern of the mayonnaise that's important; the outline of the burger is just the way the bee returns to start the dance again, circling back around one side and then the other. The angle between the mayonnaise and the vertical provides information about the direction of the food source. This angle is the same as that between the sun and the direction the dancer is encouraging her nestmates to go. Astonishingly, a bee that has been dancing for a long time will vary this angle to correct for the fact that the sun changes position as time passes. The steps of the waggle dance also tell the departing bees how far they have to go: one hundred metres for every seventy-five milliseconds spent waggling. This is a remarkably accurate system for transferring information, but how could it have evolved, given that the bees are dancing inside a log in the dark? The answer comes from studies of related honey bee species, like the giant honey bee. These studies suggest that the ancestral honey bee nested out in the open, rather than in a hive, making it much easier for the workers to interpret the waggle dance because the bee giving the directions could point directly at the food source.

Solitary bees, of course, do not need to communicate much at all. They must coordinate mating, and they may need to express nest ownership if another bee gets confused as to whose nest is whose (or is trying to usurp a superior home on purpose). These bees would get a failing grade for their communication skills.

In honey bees, stingless bees and bumble bees, the hive is a highly structured place. There are areas where new brood are produced, and places where food is stored; there is a garbage dump (usually outside the hive) that may also function as a mortuary. Within this city-like organization, there are complex social divisions, with individual bees taking on different tasks. Some perform daycare services, looking after the developing brood. Others are foragers that collect pollen and nectar. Often, an individual worker will start her adulthood as a childminder and end it foraging (literally shopping until she drops). These tasks require large numbers of individuals in a honey bee hive, just as they do in human societies. And like human societies, the hives need other specialized performers as well. Among these are undertakers to dispose of the dead and police to ensure that the other workers stay in line. But such complexity cannot be exhibited by solitary bees, which make up the majority of bee species. That some bees have evolved such highly sophisticated societies while most others are entirely solitary makes this achievement even more impressive.

≈

The final misconception is that all bees are "as busy as a bee." It is true that many bees are quite busy. For example, foraging honey bees will collect pollen and nectar from early morning until early evening, every day the weather is suitable, until they die. In Arctic Canada, the land of the midnight sun, bumble bee workers may forage twenty-four hours a day until they die of fatigue. But most solitary bee females forage for only one or two weeks before stopping for good. By the time your spring tulips have finished flowering, some of your local bee species will have completed all the work they will do their entire lives. For example, those that collect pollen from sallow and willow catkins, or from those plants that grow on the forest floor (these flower before the trees come into leaf and shade the ground), will complete all the activities required to produce the next generation in perhaps one week of work during warm early spring weather. While bees may be quite "busy," it's often not for very long.

Even some social bees are quite lazy. An example of such sloth is found in the Virginia carpenter bee, a species that is commonly mistaken for a bumble bee because it has golden yellow hairs at the front and is blackish at the back (where it is largely bald). It is called a carpenter bee because it chews holes in outdoor wooden structures, including patio furniture, grape arbours, fence posts and dead trees. Why these bees are so lazy has been investigated by the evolutionary biologist Miriam Richards at Brock University in St. Catharines, Ontario.

Miriam was my first graduate student, and now she has graduate students of her own (which kind of makes me an academic grandfather). Miriam and her students have been studying the Virginia carpenter bee right outside their laboratory on the Brock University campus. The bees actually nest in the wooden campus benches, so rather than sweating their way through the middle of some desert or getting lost in an impenetrable jungle, the intrepid researchers can just take their morning coffee outside, sit on a bench in the sunshine and gather data. That sounds easy. Alas, these bees can live for two years, so the researchers have to sit on the benches and drink coffee throughout the spring and summer for years on end. But their patience was rewarded by the interesting discovery that the lazy females in a social nest were actually interloping home-owners-in-waiting; they would sit in a nest all day for over twelve months waiting for the owner to die so they could inherit the nest. These patient but lazy females were mostly unrelated to the nest's rightful owners. It's as if some stranger invaded your home, waited for you to pass on to the big bee's nest in the sky and then claimed squatter's rights. Given that to build their nests, these bees must chew through tough wood for a distance that would, for us, be the equivalent of perhaps one hundred metres, it is understandable that some females would rather be patient interlopers than spend an enormous amount of time wearing away their teeth to construct a home for their offspring.

What is even more surprising is that there are well over two thousand species of bees that do not construct nests or collect pollen at all—these are the cuckoo and socially parasitic bees, which lay their eggs in the nests of other bee species. Female cuckoo bees are busy flying around looking for nest sites, but they are not busy in the typical bee-busy way (foraging for pollen and nectar, constructing a nest or looking after the young). Cuckoo-like behaviour has evolved independently among bees on numerous occasions (we are not yet sure how many times, but probably over a dozen). Most of these bees invade the nest of a solitary host while the owner is away foraging.

The social parasites have a more difficult task. They have to invade the nest of a social species, which will normally have at least one adult at the ready for defence. Once inside the nest, these invaders suppress or kill the queen and replace her as the primary egg-layer. The workers of the host bee now slave away to raise the offspring of a completely different species.

When it comes to aptitudes for hard work, we again have a wide range of variation. The Arctic bumble bees that work constantly until they die of exhaustion get 100 on the busyness scale. At the other extreme are the cuckoo bees, which collect no food and dig no nests and therefore score zero on the busyness scale. Most bees fall somewhere between these two extremes.

So now we know that only female bees sting, and only some females at that. We know that far from being hard-working, many bees are rather inactive, and some are even social parasites. We know that only a small proportion of species have complex social lives, that few bees live in hives and that almost none of them makes honey. So what, then, are bees? As we'll see in the next chapter, this is not an easy question to answer.

4

A Bee or Not a Bee? A Difficult Question to Answer

INSULTING THE EXPERTS IN PORTAL, ARIZONA

It was my first lecture in the annual International Bee Course, and I was about to make myself very unpopular with the other instructors.

The International Bee Course is an intensive ten-day workshop held every August or September at the American Museum of Natural History's southwestern research station in Portal, Arizona, about three hours east of Tucson. Students come from all over the world to be taught by experts on different groups of bees. My primary role was to explain the morphology of bees and to give an overview of one particular group, those small, often sociable sweat bees. The first time I helped teach the course, several students commented that while they had learned to identify different bees, they had never been told how to tell bees apart from other insects. So the next year, I decided to administer a quick test to see how good the attendees were at making the distinction.

I had prepared a PowerPoint presentation showing ten images: five were of bees and five were of insects that look like bees but aren't. I warned the audience that they would have only five seconds to

51

decide if they were looking at a bee or not. Everyone at the workshop, instructor and student alike, was given an answer sheet. I allowed them to answer the questions anonymously, but I did divide them into three groups: students who had previous experience identifying bees, students who were complete novices and the other six instructors. I wanted to find out whether a little knowledge was a dangerous thing, and whether a lot of knowledge actually made a difference.

Slide one showed a very waspy-looking moth (this one was easy). Slide two featured a beautiful tricoloured bee that looks very much like a bumble bee but is actually a rare cuckoo bee from Thailand. Slide three was of an almost bald bee, with black-and-white markings at the front and orange at the back; it sure looked like a wasp. Slide four was of a very annoying fly (the larvae of this species had been eating my flower bulbs all summer) that mimics a bumble bee. Slide five showed a wasp that looked remarkably like the bee in slide three. Slide six depicted a very odd Australian bee that looks incredibly un-beelike (it's almost hairless, with tiny eyes and large protruding mandibles). Slides seven and eight were of bees that look very much like yellowjackets. The remaining two slides showed a bee-like beetle with enormously hairy hind legs that look tailor-made for collecting pollen and a close-up of a wasp with its tongue extruded. This species has a tongue that looks very much like the tongue of some bees.

I saw glum looks from the instructors, but nobody seemed really unhappy. We had to wait for a tallying of the results before the complaints would start.

Ten minutes later, we had those results. To my surprise and the instructors' horror, the "experts" were right, on average, only 75 percent of the time, while the newcomers and those students with some experience studying bees were both correct approximately

65 percent of the time. There were only three images that all the instructors got correct—the moth, the fly and the beetle (all of these, although quite bee-like, were comparatively easy for those with a lot of experience identifying insects to recognize). The other non-bees—all wasps—were incorrectly identified by one or more of the instructors. Most of the bees were correctly identified by the instructors, though slide six—the small broad-tongued bee from Leyburn—fooled all but one of them. It's a rare bee with an unusual cylindrical body that enables it to nest in abandoned beetle burrows. Some instructors were even certain they knew which group of non-bee insects this odd animal belonged to.

Of course, the instructors were somewhat embarrassed to have made so many mistakes, and some complained that they didn't have enough time to look at the key characteristics. But I countered that, in the field, they would never have five seconds to decide whether something was a bee or not. I did not add insult to injury by pointing out that the images provided were enormous—more than a hundred times as large as most bees and wasps. In fact, what the experts really are expert at is deciding whether a small, life-size insect is a bee while it flies around. This is actually somewhat easier than trying to identify static images because an organism's behaviour can be as important as its appearance in aiding identification. There are some bee species I can more easily identify with a live individual in front of me than with a pinned one under the microscope. There is something about the gestalt of a live individual seen with the naked eye that is not so detectable when the same organism is impaled on a pin and looked at using high magnification.

My test was, of course, a little misleading. I had chosen the most bee-like non-bees I could find, and most of the bees I had chosen were astonishingly un-beelike and very obscure ones. Nonetheless, it did rather nicely illustrate the point that bees are not easy to tell apart from other insects. So how do we identify bees? To answer this, we have to go back in time to the early Cretaceous period, over one hundred million years ago.

Imagine you have travelled back to the time of the dinosaurs on a field trip arranged by some futuristic university department of empirical palaeontology. Grass has not yet evolved, and some dinosaurs are wearing feathers. Flowers are few and far between. Our forebears are tiny shrew-like things trying to avoid being trodden by large dinosaurs or eaten by smaller ones. They have no time to pay attention to one of the major events in evolutionary history: somewhere out there, a species of wasp is evolving the habit of foraging for pollen rather than catching other insects as food for its offspring. This is the start of something important, a dietary revolution that will eventually bring the world numerous spectacularly colourful, often wondrously aromatic flowers. It will also bring the heavenly elixir of honey, the complexity of the social lives of honey bees and the marvellous attention to motherly duty and self-sufficiency of the numerous solitary bees, all of which, I will soon argue, can help us understand the state of our human-modified environment. One hundred million years later, we have well over nineteen thousand species of bees—and those are just the ones we've recognized and described so far.

The closest relatives of the bees are the apoid wasps, from which bees evolved. Most apoid wasps construct nests in the ground, as most bees do. Others live in plant stems, and some, such as the mud daubers, make earthen chambers on walls or tree trunks. Some apoid wasps hunt caterpillars, paralyzing them and dragging them to their nests, where they will lay an egg on their victims. The wasp larva then eats the juicy, fresh food that its mother has provided for it.

Others hunt cockroaches, grasshoppers or flies, which aren't as juicy but are good sources of protein nonetheless. We have already met the honey bee wolf, a member of a group of over thirty species of digger wasps that hunt bees. Honey bee wolves also use honey bees as a source of energy for themselves. They will grab a honey bee, paralyze it and squeeze the nectar out of its stomach. (Presumably they paralyze the bee so it won't fight back while having its last meal forcibly regurgitated.) When finished, the honey bee wolf will throw its empty victim away and, if it's still thirsty, go catch another one.

Some digger wasps go for smaller prey, such as thrips or aphids. The aphid-collecting lifestyle makes good sense as an evolutionary intermediate between wasp carnivory and bee vegetarianism; after all, aphids secrete honeydew, the shiny, sticky stuff that gathers underneath their colonies. They produce honeydew because their food is so lacking in amino acids (the building blocks for protein) that they have to consume far more sugar than they need. The excess is excreted as honeydew in amounts that can be prodigious.

I suspect that the ancestral bee may have developed a taste for small, sweet food items from an evolutionary history of collecting aphids for its young. But whether aphid-eating was an intermediary stage between digger wasp and bee or not, the apoid wasps were the progenitors of bees. This means that, evolutionarily, bees are digger wasps that went down the food chain to collect pollen rather than other insects for their young to eat. This was a major evolutionary transition with massive ecological repercussions.

~

The worst example of bee misidentification I know of can be found in one particular edition of a book entitled *Bees of the World*, which had a fly on the cover. A similar lapse occurs on a book called *A World without Bees*—its cover suggests that such a world would also be without hover flies. My university once ran a story about the research of one of my students and illustrated it with a picture of a wasp (a type of wasp that was as close to a bee as one can get, but still not a bee). Similarly, if you peruse images of insects on the web, you will find many bees that are identified as wasps or something else, and many wasps and other insects that are identified as bees.

Because of their waspy ancestry, many bees are difficult to tell apart from their carnivorous relatives. This is particularly the case for the less hairy bees, such as those that carry food in their gut instead of on pollen-carrying hairs or those that do not collect pollen at all but instead lay eggs in the nests of other bees. The most commonly observed bees that fit the former description are the masked bees. These are usually black insects with white markings, particularly on the face and the legs. The females typically have a white or cream-coloured triangular spot on each side of the face, whereas the males' faces are usually entirely pale, at least beneath the antennae. These colours are not formed by patches of pale hairs on a dark background, but instead are the colours of the external skeleton. Found worldwide, masked bees are small bees that generally nest in stems. If you look into the hollowed-out end of an old raspberry cane in the late afternoon of a summer day and see a small white or cream-coloured face looking up at you, it is likely a male masked bee getting ready to spend the night.

The other wasp-like bees are the cuckoo bees, and they are very diverse, partly because the cuckoo bee lifestyle has evolved independently numerous times. The most common cuckoo bees in temperate regions of the northern hemisphere are the nomad bees, which are most easily found flying around earthen banks in spring as the females look for the nests of their solitary mining bee hosts. Nomad bees are brightly coloured, and again, the colours are on the body surface and do not result from patterns of hairs. Some are black with orange markings on the head, thorax, legs and abdomen. Others really look like small yellowjackets, with yellow bands on a black background. Still others are mostly orange or red with yellow bands or spots, or have bands of yellow and brown. These are all very waspy features, but these organisms are certainly bees.

So how do you tell a bee from other bee-like insects? The simplest answer is that bees have branched hairs on their bodies. This doesn't help much, given that most bees are tiny insects—the smallest are 1.6 millimetres long, and six hundred of them can live together in a nest the size of a walnut. These bees are so small it is difficult to tell whether they have hairs at all, let alone hairs that have branches. To see the hairs, you need a microscope, and either the bee has to be dead or you have to hold it still long enough to look at it with high magnification. To complicate matters, many bees have some hairs that are not branched, so you may have to look quite hard to find the ones that are. Still, it is uncommon for a bee to have so few branched hairs that you cannot find them relatively easily, at least with high magnification.

There are a couple of other features that enable us to tell bees apart from other insects. Bees have a flattened metabasitarsus (a comparatively apical segment of the hind leg) and the females have a spiracle plate that is completely divided. (The spiracle plate is the most lateral part of the sting apparatus.) Wasps, on the other hand, have a cylindrical hind basitarsus, and their two spiracle plates are joined together by a hard, dark skeletal rod.

Unfortunately, checking whether the hind leg is flattened or cylindrical is not much easier than checking for branches on the hairs. Although the spiracle plate can be seen easily in a large bee, anyone who would pare apart the tail end of a large, potentially stinging insect to see if it is a bee or a wasp is destined to have an uncomfortable life. Even with a microscope, these features are not easy to see in a moderately sized specimen, and they're impossible to see if the insect is flying around, regardless of how large it is. So how can you tell whether something that is alive and outdoors is a bee or not?

Any insect that you see collecting pollen and storing it on its hind legs or on the underside of its abdomen will be a bee—a female bee collecting food for its offspring or for its younger siblings if it is a worker of a social bee species. In the absence of a microscope and previously collected specimens for comparison, you can tell male bees through their association with their females. Males will usually be of a similar colour, and they are narrower and have longer antennae. Like male humans, male bees have more hair on their faces than females do, but unlike humans, female bees have more hair on their legs or abdomen because that's where they carry the pollen. Cuckoo bees can best be identified as they fly around the nests of their hosts. If you find a nest site for a mining bee and see some wasp-like insects flying around looking in the entrance, there is a good chance that they will be cuckoo bees. In most of the world, the only bees that bring pollen and nectar back home in their digestive system rather than on their legs or abdomen are the masked bees. (Australia is an exception. Numerous species unique to that continent transport food internally. These are the broad-tongued bees, such as the one from Leyburn that fooled all but one melittologist in the bee quiz I mentioned earlier.)

I have explained why many bees look nothing like the typical bee. But why do some insects that are not bees look so bee-like?

If you have ever been stung by a bee, you are probably not keen to repeat the experience. Imagine what it would be like for a small bird or a small mammal to receive the same amount of venom in its tongue or the roof of its mouth. What would happen to a parent bird with a clutch of hungry nestlings all squawking for food? Having attempted to catch a bee to feed to them, the stung parent would likely be incapable of catching anything else for a long time. There is a good chance its brood would starve to death while the parent recovered. Or perhaps the amount of venom received would be enough to kill the parent too. Natural selection would favour an insectivorous bird or mammal that learned to avoid stinging experiences. Natural selection would favour a parent bird that avoided even the learning experience. The bright colours of many bees and wasps serve as a powerful reminder to predators that organisms with these warning colours are to be avoided. That gives any other organism that looks similar an advantage. This explains why there are many insects that are not bees, are not even closely related to bees, but look so much like them: they have evolved bee-like characteristics because stinging bees are left alone by predators. One of my colleagues has referred to such insects as "wanna-bees."

There are many beetles, moths and flies (such as the one whose maggots eat my flower bulbs) that have evolved to bear an uncanny resemblance to some bees. Once an insect has started on the evolutionary path of mimetic resemblance, natural selection will tend towards perfecting the similarity—any new mutation that improves the resemblance will fool more predators, and individuals that more forcefully evoke past memories of being stung will avoid getting eaten.

But what of bees that do not sting? Interestingly enough, some of these have evolved to mimic bees that do sting. I remember sampling bees from the flowers of a wild relative of tomatoes in northern Argentina. They flew rapidly into a flower, buzzed loudly to extract pollen from the anthers (like blueberries, tomatoes have to be "buzz-pollinated") and then flew on to the next one. They were beautiful large bees, blackish brown at the front with a lovely silky red abdomen. They were a type of mining bee I had not seen before in the wild: the red southern oxaea. Having just finished a very long and rather tedious manuscript on bee-sting morphology, I knew that none of the bees in the oxaea group could sting; I had devoted several pages to the topic of the reduction in the sting-related structures in these and other bees. Safe in the knowledge that these bees could not sting me, I decided to catch a few, taking them out of my net between finger and thumb. All of a sudden, I knew something was wrong. Some bees have such a large sting shaft that you can feel it penetrate the skin even before any venom is injected. Ouch! I had been fooled; this bee wasn't what I thought it was. My flesh had been penetrated by the sting of a ruddy centris. I knew from my aforementioned research on stings, and from previous painful experience, that these bees had a long sting. That individual ruddy centris got away: its sting had served its function perfectly, and its bearer got away with its life after having been caught by a predator (me). This is not something that I often let happen, even with bees that sting me. When I get stung, I can usually control the natural reflex to release the offending insect. Because I am interested in catching the bee and am already aware of the possibility of pain, I am normally psychologically prepared for it when I pick up the bee.

In this instance, a bee without a functional sting had evolved to resemble one with a fierce sting. We know that this was the direction of change in colour pattern because none of the relatives of the red southern oxaea is black and red (some are all orange-brown, some are all black, some are dark brown with spectacular metallic bands of bright green and some are bright green almost all over). In contrast, there are numerous stinging bee species like the ruddy centris that are black and red, common warning colours among insects. A bee without a sting had evolved to look like a nasty stinging species that lived in the same part of the world and also foraged for pollen on exactly the same plants. I promised myself to look more carefully in the future at the big bees in my net before picking them out with my fingers.

As this example demonstrated, completely unrelated bees can often look so similar that someone with years of practice can have trouble telling them apart. The situation is much more difficult when we try to tell *closely* related bees from one another. In the next chapter, we will find out how biologists do this.

5

Two Bees or Not Two Bees?
An Even *More* Difficult Question to Answer

UNCOVERING IRREGULARITIES AT THE NATURAL HISTORY MUSEUM,

LONDON

In August 1978, I spent a lot of time sitting at a microscope in the prestigious halls of the British Museum of Natural History, as it was then called. (Now it is rather egotistically called the Natural History Museum, as if there are no other natural history museums anywhere else in the world.) Gaining access to these hallowed halls is not easy, but with the right contacts, the serious student can eventually obtain permission to work on the collections behind the scenes. My reason for this visit was to research a wasp that I could not identify; it did not match any of the species known in Britain.

At most museums, the vast majority of holdings are off limits to the public. Recent trends in museology promote opening up the restricted areas by installing glass walls between the researchers and the public. This creates a somewhat zoo-like atmosphere, where the scientists are on display—and visitors are discouraged from feeding the animals. (There is usually a cafeteria for them somewhere in the basement.)

It is difficult to overestimate the impact of the deep sense of history that permeates the musty air in institutions as venerable as the Natural History Museum. Enormous antique wooden cabinets extend from floor to ceiling. These are storehouses of the scientific knowledge gained over more than two centuries of study—repositories of the painstaking activities of generations of people who dedicated their professional lives to amassing and describing the diversity of the natural world.

This was tireless and rather lonely work, and perhaps as a result these folks were often idiosyncratic. One entomologist, for example, was notorious for ramping up the rate at which he discovered new species each time he and his wife had an additional child. (Back then, a scientist's pay depended upon the number of new species described.) It was said about another that it was easy to tell which illustrations he drew before and after a liquid lunch. Even the toilet paper was unique at the Natural History Museum: almost until the end of the twentieth century, every single sheet of it was stamped "Property of Her Majesty's Government" in purple ink.

Within each drawer of the museum's wooden cabinets, hundreds of specimens are pinned on slats of balsa wood. Each pin bears the specimen itself and several tiny handwritten labels. (Handwriting experts would have a field day with such collections. I imagine they would readily detect fastidious personality types and obsessive-compulsive disorders.) A typical label will list where the specimen was collected, the date the insect met its death at the hands of the entomologist and the name of the perpetrator of that deed. Another may give the species name and the name of the person who identified it. Species names are also written in a larger, more flamboyant hand, on larger pieces of paper or white card at the head of the column of specimens. Everything beneath that label is supposed to be that named species until another large label announces a new species. However, you cannot assume that everything between one label and the next belongs to the first species mentioned. The first time I discovered a case of mistaken identity within these hallowed halls I was shocked. I couldn't believe it! How could anyone mistake a white-banded sweat bee for a banded sweat bee—and at *the* Natural History Museum, no less? Since then, I've become more accustomed to finding examples of mistaken identity in even the world's best museums, few of which have a resident bee identifier to keep the collection error-free.

It is a nerve-racking experience to be given access to the inside of these drawers for the first time. Extracting an insect from the side of a drawer is comparatively easy, but accessing one from the middle is a difficult task because of the rows and rows of specimens surrounding it. Visitors with long sleeves are not encouraged to look inside these drawers. Dragging the cuff of a Victorian cloak or a modern sweater across a row of century-old specimens can send insect parts flying off in all directions. Such accidents may result in instantaneous expulsion from the premises. Most drawers bear evidence of past blunders, with some specimens headless or missing the abdomen, and a smattering of entomological body parts lodged in crevices here and there.

Like all endeavours, insect identification is something that becomes easier with practice. Correctly naming the first specimen may take hours, but once you memorize the features of that species, you will be able to identify additional specimens instantly. With experience, the gestalt of a species becomes lodged in the cranium, making it possible for an insect to be identified even as it flies by. But obtaining that experience takes a long time, and ultimately you must rely on the activities of all previous researchers who have performed similar work.

How is it that melittologists decide which species of bee they have? Indeed, what does the term "species" even mean? Given that species are the most fundamental units in all of biology, it is perhaps somewhat disheartening to learn that biologists argue, often extremely heatedly, over how to define the term. (There are almost as many definitions of the word "species" as there are people interested in debating the issue.) I will take the middle ground amid the many battling sides and suggest that different species can generally be told apart by characteristics that are inherited (suntans don't count but DNA sequence differences—when large enough—do), and that these differences usually persist among species at least partly because they are reproductively isolated (in other words, different species do not usually interbreed). Most people who perform species identifications—that is, taxonomists—do not play around with living organisms, so it is the first part of my definition that is most relevant. Taxonomists have traditionally spent their entire professional lives looking for morphological distinctions among specimens and evaluating whether the differences are sufficiently large for them to decide that they have discovered, or confirmed, species-level differentiation.

Taxonomy is the branch of science that deals with the identification of species, and it has a long history—practices that are still in use today can largely be dated back to the Swedish botanist Carolus Linnaeus in the mid-1700s. Linnaeus was an interesting fellow: while concocting his Latin-based nomenclatural system he even Latinized his own name (he was born Carl von Linné). Since then, scientists have developed a complex set of procedures to increase the probability of correct naming. These, to most people, can seem mind-numbingly pedantic. The regulations associated with the naming of animals, for example, are crystallized (indeed, almost fossilized) in the International Code of Zoological Nomenclature (ICZN), and the book that outlines these rules is 126 pages long and reads like a legal text. If you have insomnia, I can recommend the ICZN regulations. (If you find them too stimulating, there is a similar tome that outlines the procedures for naming plants and another for naming bacteria.)

Although these guidelines do not make exciting reading, they are absolutely essential to the human enterprise (even though the only people who realize this are the taxonomists who use them and the few enlightened students of other aspects of biology who value the identifications taxonomists provide). Taxonomy is a surprisingly important endeavour. Pests inflict trillions of dollars worth of economic and medical damage every year, and errors associated with misidentification have proven extremely costly. There are cases, for example, where the wrong species of biological control organism has been introduced, permitting the pests it was meant to control to increase in both number and economic impact. Sometimes even the pests themselves are incorrectly identified. Such inaccuracies can extend to disease-causing or parasite-transmitting organisms. Even the accurate identification of an illness follows similar procedures. A doctor once misdiagnosed my older daughter's salmonella as "rice" (he must have had his microscope the wrong way round). All these errors can be traced back to an inappropriate application of the identification routines codified in the books that outline taxonomic procedures.

So how is it that taxonomists put a name to something? Well, this process is beginning to undergo a massive change. Rugged individualists working away at museum benches are being assisted by small robots that automatically generate short DNA sequences called DNA barcodes. But even this newfangled method ultimately relies upon identifications that are made just as they were in the times of Linnaeus—that is, through comparative morphological examination and comparison of specimens.

Every correct morphological identification is ultimately made in reference to a type specimen. This is tantamount to an application of Plato's theory of forms. Plato's view was that all objects are inferior representations of an archetypal example of that object. For example, the laptop I am typing on is but a poor representation of a perfect laptop that exists somewhere in the ether.

Of course, nobody today would suggest that there is a perfect laptop somewhere in the universe, and that all laptops here on earth are imperfect representations of it. Indeed, taxonomy is the only area of human activity I know of where something akin to the platonic theory of forms applies. The term taxonomists give to these representative examples is *type specimen*—or more precisely, *holotype* and, for sexually reproducing organisms, *allotype*. This additional term is needed for organisms with separate sexes because males and females of the same species often look completely different. We call the example of one sex the holotype and the example of the other sex the allotype. Which way round these terms are used is up to the person doing the describing.

Ultimately, all identifications have to be made in reference to these ideal forms, and the museums of the world hold millions of them (at least one for each described species). Because type specimens are the ultimate name arbiters, they are kept under lock and key and are usually housed in separate cabinets away from the main collection. Alas, some entire collections were destroyed during the Second World War. The Natural History Museum moved most of its important holdings to a barn in the countryside to avoid destruction (since London was a prime target of Hitler's bombs). But imagine my surprise when I looked at one particular type specimen and discovered that this long-dead insect had had its head knocked off and then re-affixed to its body with an enormous glob of glue. As if that was not bad enough, the head had been replaced backwards. This was unfortunate because, like people, bees have many more useful features for identification on their faces than they do on the backs of their heads. Other type specimens have been found entirely lacking their heads. Recently, one of my students found a headless type specimen in a collection—but he found the missing body part stuck underneath the plastic foam that lined the specimen box. Sometimes the so-called perfect form is an absolute abomination that would have given Plato nightmares.

What these arcane taxonomic procedures actually mean in practice is that the identification of any specimen involves reference back to the holotype of that species—even if it has had its head reattached the wrong way around or only a few fragments of it remain. But when even worse things than decapitation happen—whole collections lost in warfare, destroyed by museum beetles or damaged in a fire—there are recommendations on what to do. For example, if a holotype is lost, it is permissible for an acknowledged expert to decide which other specimen of that species can serve as the replacement ideal form. Put another way, the ethereal perfect laptop can be replaced by another previously less perfect one, but only with the say-so of someone with impeccable qualifications.

As we will find out later, scientists conducting a bee-biodiversity survey can often collect thousands of specimens in one field season. It would not be feasible to compare each individual to a holotype even if you had all these entomologically precious specimens at your fingertips. Under these circumstances, the researcher does the best he or she can, consulting published identification guides and comparing the specimen to more broadly available ones that have actually been compared to the type specimen or to those that were available to the original describer. (These latter specimens are called *paratypes*, literally "next to the type," which is where they likely were in the tray while the taxonomist was working.) With these and other shortcuts, researchers can be moderately confident that they have correctly identified their study organisms. Synoptic collections that represent a particular local fauna are especially useful and, of course, specimens can be compared to type material when there is doubt, but if an identification guide is adequate, individuals can be named without so much time-consuming comparison—sometimes.

Identification guides published as books or in reputable scientific journals are the tools most people use to identify bees or other tricky critters. The key to the 480 or so genera of bees of the world itself forms a book that is almost one thousand pages long. Other keys may permit identification of the genera of specific regions. There are also keys to species of particular genera, although these are usually hidden away in obscure research journals. The number of countries for which species-level keys are available for the entire bee fauna is small. New Zealand and Madagascar are two of them. Islands have fewer species than mainland areas because they are difficult to immigrate to and the habitat area is less extensive. Populations are also more likely to get wiped out on an island, and the chances of them recolonizing from a nearby mainland are low. New Zealand has 38 species of bees, compared to over 2,000 for Australia. Madagascar has 238, compared to more than 1,000 for East Africa.

With the advent of the Internet, numerous tools have been developed to facilitate identification. Digital images can be used to illustrate the characteristics in the key, and whole-animal images can be taken of the males and females of each species. But even with these modern efforts, the language used is often archaic and user-unfriendly, particularly to the uninitiated. For example, in common parlance we would never say that a bald person has a glabrous head. But if a part of a bee that is normally hairy is bald in one particular species, the key will probably describe it as glabrous. Or imagine going up to people with red hair and telling them that they have ferruginous pubescence. You might get yourself into trouble! But this is exactly the phrase you will find in identification guides to refer to species that are red-haired.

Another problem is that keys are written by people who don't need them for people who can't use them. Those who write the key know the organisms they study so well that they do not need to use the results of their own work. Conversely, those who need to use the key generally can't interpret it accurately. Many keys have the word "usually" as a descriptor. This is useful only if you have a "usual" specimen. But the beginner has no idea if something is usual, unusual or just plain weird.

Some keys differentiate bee species partly by the floral host from which the specimen was collected. But what if you caught one because it flew in the window of your car while you were driving? What if your botanical knowledge is as poor as mine? What if the specimen came from some kind of insect trap?

These morphology-based approaches have been standard practice for centuries, but sometimes they do not work even for well-known species being studied by would-be experts, including me, as the following story demonstrates.

For my Ph.D. at the University of Toronto, I compared the social behaviour of what I thought were different populations of the ligated gregarious bee. This species was easy to identify—it's dark

brown and has narrow white bands on an abdomen that tapers mark-
edly towards the waist (hence its ligatured appearance). That could
describe several hundred bee species, but what makes this one particularly easily recognizable is its head, which is as wide as or wider
than the thorax and almost triangular in side view. It was considered
one of the most widespread bee species in the world, and could be
found from southern Canada and the United States on into Mexico,
through Central America, and into Colombia and the larger Caribbean islands. In 1994, long after my thesis was completed, we realized that this easily identified species was in fact two different species.
Just last year we obtained some information that the Mexican populations might actually be a third distinct species. All these species are
morphologically inseparable but genetically well differentiated. In
the southern foothills of the Appalachian Mountains, two of the species can be found nesting in the same patch of soil and foraging on
the same flowers, but they do not interbreed and so remain genetically distinct. What was previously a single, easily identified species
is now a complex of perhaps three (or maybe more—we do not yet
have data from Central America or the Caribbean) "cryptic species."
These are species that are extremely difficult, if not impossible, to tell
apart using traditional morphological approaches. The discovery of
such cryptic species happens often, but only when people spend a
long time looking.

    These findings are somewhat worrisome. After all, if tradition-
ally trained taxonomists cannot always differentiate among genetically isolated species, then how can the average biologist find out
whether it is "two bees or not two bees," or even three or more
bees? Fortunately, scientists have developed a new method—DNA
barcoding—that promises to automate the identification of almost
all species.

∽

The Canadian biologist Paul Hebert of the Biodiversity Institute of Ontario at the University of Guelph invented DNA barcoding. Paul is highly energetic; he's the only person I know who will regularly send out emails well after ten o'clock at night and then start again soon after four the next morning. Despite the extremely long hours, he manages to work intelligently and with good humour for all of them. As a result, he has obtained numerous multi-million-dollar research grants and built up a large research team, all with the goal of providing automated species identification through the use of DNA barcodes.

First described in 2003, DNA barcoding has already been proven to have numerous practical applications, such as distinguishing different disease-vectoring mosquitoes, identifying potential invading species that arrive in shipments of imports from far-flung countries and finding out whether a fishmonger's purported red snapper is indeed red snapper and not an endangered fish species or poisonous fugu.

The way DNA barcoding works is remarkably simple. A small segment of DNA from a tiny piece of tissue is amplified using complicated chemical cocktails that are designed to recognize and copy one short sequence. This one portion is copied thousands of times, and this rich soup is then sent to a machine that reads the genetic code that is the most abundant ingredient. The machine generates a barcode like the ones you find on merchandise in your local store, but instead of being black-and-white, it comes in four different colours: red, green, blue and black. Each colour represents one of the four letters in DNA's genetic blueprint.

DNA is the basis of inheritance, and it varies both within and among species. So the choice of which fragment to use for automated identifications is important. The chosen segment has to have more signal than noise. This means that the level of differentiation among species has to be greater than the variation within a species, otherwise the sequences would not be diagnostic. In other words, there have to be more differences among the sequences of closely related species than there are between individuals of the same species. The fragment that is best suited for this purpose for almost all animals is a small part of the mitochondrial DNA (mtDNA for short). The mitochondria are the minute powerhouses of our cells, and there are numerous mitochondria in each one. (This contrasts with nuclear DNA; because there is only one nucleus in the cells in an organism's body, nuclear DNA is much less suitable.) Because there are numerous mitochondria in each cell, a chemical cocktail aiming for mtDNA will have many more targets to hit. Another advantage is that mtDNA is a circular molecule that is inherited as a unit and passed on to the next generation only through the mother. Because the whole sequence is passed on to each offspring intact, any mutations originating in one maternal line cannot get mixed up with different mutations arising in another lineage. (A second mutation in the same lineage will always be associated with the first mutation. A third mutation in the same lineage will always be associated with the first two, and so on. All subsequent mutations will always be associated with the first one.)

In contrast, mutations arising in nuclear DNA are shuffled around from one generation to the next as a result of sexual reproduction. Mitochondria aren't involved in sex, which makes tracing the history of evolutionary lineages easier using mtDNA. Deciding which species an individual sequence comes from is also easier.

The long-term objective of the barcoding enterprise is to obtain DNA barcodes for all the species on the planet. Not surprisingly for a technology that is only seven years old, we are still at the data-acquisition phase with bees and most other groups of animals: approximately 18 percent of the world's bee species have had one or more specimens barcoded and the sequences put into the database; that's quite an achievement in such a short time. While taxonomists like me toil away at identifying specimens from which the database can be constructed, other researchers are working at miniaturizing the components that actually do the barcoding. It should soon be possible to make identifications using a hand-held device. This cellphone-size barcoder will perform the amplification and sequencing phases of the process and send the information to a centralized database through the Internet. A central computer will then reply with a species name. This may sound rather far-fetched, but there are already devices for DNA extraction and amplification that are small enough to be held in the hand. Similarly, small battery-powered options are becoming available for DNA-sequencing machines as well. The future has an interesting habit of speeding up on us. Most developments happen faster than even futurists expect them to. Current guesstimates are that hand-held, cellphone-size DNA barcoders might be available in ten years. It will be interesting to see when they actually arrive. Alas, recent funding cuts by the Canadian government have slowed progress.

Nevertheless, DNA barcoding is already living up to its promise of revolutionizing our understanding of the diversity of biological treasures of our planet. In one national park in Costa Rica, the renowned ecologist Dan Janzen has a small army of assistants who scour the forest for caterpillars, which they then take back to the laboratory and rear to adulthood. This way the researchers can associate the caterpillar (which is generally unidentifiable to the species level because there are so few specimens of accurately identified insect larvae in museum collections) with the adult moth or butterfly (of which there are millions of named specimens worldwide). Perhaps not surprisingly, given the complexity of the data he was obtaining, Dan Janzen was an early convert to DNA barcoding, and some of the most exciting examples of its application involve collaborations between him and the group at Guelph.

For example, they have found that what was thought to be one species of butterfly is actually at least ten different species. Not only are the species genetically different, but their caterpillars have distinct colour patterns and are usually associated with different foodplants. The original one "species," known from Mexico to Argentina, turns into ten different species in just one Costa Rican national park; heaven knows how many additional species will eventually be found throughout the rest of Central and South America. And that's just the butterflies. The situation seems even more complex for the parasitic wasps and flies that emerge after eating a caterpillar's innards. Some of these parasites eat the insides of only a single host species; others are less fussy and eat the internal organs of a wide variety of different grubs. At least that was what was generally thought. DNA barcoding of the apparently less fussy parasites has shown that one species of generalist may actually be more than a dozen species of specialists.

In a nutshell, DNA barcoding is revealing that nature is more diverse than previously thought—in some cases, an order of magnitude more diverse. With the bees we seem to be more fortunate, perhaps because bee taxonomists have long been diligent in looking for differences among species. In most of the groups of bees studied so far, the increase in number of species is modest, and I doubt that the actual total revealed by barcoding will be more than 50 percent above current estimates. Nonetheless, that's a substantial increase—from approaching twenty thousand species to approaching thirty thousand.

Unsurprisingly for a development that promises to let anyone do something that was previously the domain of only a select few experts, DNA barcoding has been controversial. Some have claimed that it does not work well enough. But when I look at the number of misidentified specimens in museums (that banded sweat bee misidentified as a white-banded sweat bee), it is obvious that DNA barcoding works much better than traditional approaches. Some have complained that DNA barcoding is expensive and will take funds away from traditional taxonomists. But the recent history of taxonomy, at least in Canada, suggests that the whole activity would likely have died out within a few decades. The number of people doing taxonomy, the number of species those people have described and the average number of grant dollars for taxonomic research have all declined precipitously in Canada since 1980. In fact, the resources that have gone to traditional taxonomists who are also involved in DNA barcoding have single-handedly taken taxonomic funding back up almost to 1980 levels (though this too suffered from the latest unfortunate budget cuts).

Perhaps the worst nightmare of some traditional taxonomists is that they will soon be replaced by DNA-sequencing robots. But this won't be the case, at least not for hundreds of years. DNA barcoding is revealing that the world is even more diverse than previously thought. All these newly discovered species have to be described by traditionally trained taxonomists. These scientists will be able to concentrate on the urgent task of describing the unknown 90 percent of the world's organisms while the routine identifications that otherwise take up so much of their time can be performed by a small army of R2D2s.

So there is an exciting revolution taking place in what was recently considered to be a dull, dusty backwater. These advances in taxonomic procedures will soon make it possible for anyone with a few dollars to spend to put a name to the organisms they find in their backyard or on a tropical vacation. Imagine the excitement of discovering that the wasp in your backyard has not yet been described! (If the wasp is a small one, the chances of this happening are remarkably high.) If you collect a few individuals and send them to an expert, he or she may even name the species after you.

These are the ways that people go about identifying organisms. Knowing the name of something is an essential step in finding out about its biology. In the next chapter, we will learn about the bees' diverse ways of being.

6

It's a Bee's Life

UP BEFORE DAWN IN THE AUSTRALIAN OUTBACK

Crunch, crunch, crunch, crunch. It sounded like marching jackboots from a Second World War movie. Although I was hearing the footsteps of only one pair of boots outside my tent, the sound struck terror all the same. Long before dawn, it was time to get up to study bees, and I am normally a late riser.

I was on a field trip arranged by Mike Schwarz, an expert on wild bees based at Flinders University in South Australia. Mike has been studying bees since he was a high school student. We had spent the day before driving from Adelaide to Lake Gillies, listening all the way to a test match on the radio—my, how the Australians love their cricket. I have never liked cricket and used to find it boring even to play, but it was surprising how infectious the research team's enthusiasm was.

Arriving late in the afternoon, we set up the tents, cooked a meal and retired early for the night because of the rigours of our morning start. Now, at 4 a.m., it was pitch-black except for the light that came from the campfire Pam Hurst had lit to start breakfast. Like most scientists, Pam is a martyr to her research. She had got up even earlier to ensure that the rest of us woke up to the aroma of fresh coffee (about the only thing that could encourage me to stir from my slumber).

Why did we have to get up so early? We were studying a social bee that nests in the dead twigs at the ends of the lower branches of the bullock bush. The project's objective was to discover just how social these bees are, so we had to find the nests before any of the bees left to start foraging for the day. This meant that we had to begin work long before the bees did and well before it got warm—and it warms up quickly in the South Australian desert, where temperatures readily reach over 40°C in the shade. Indeed, it was often too hot—and more often too dry—for the introduced, invasive honey bees to behave normally. On one particularly hot and dry day, they were so desperate for moisture they gathered around the tops of our beer bottles, putting thirsty field workers at risk of getting stung in the mouth. They also followed us into our tents because these were slightly more humid than the great Australian outback outside. We were in for a very long, quite uncomfortably hot day.

Vast amounts of predawn coffee later, we headed off, walking through the bush and attaching brightly coloured flagging tape to some of the trees we passed. Getting lost in the Australian bush is easy; getting lost before the sun comes up is even easier. This was before the advent of affordable GPS technology, and the flagging tape was to help us find our way back to the field vehicle. If any of us were later than expected returning to the car, the horn would be honked repeatedly. That same signal at any other time would indicate that someone had been bitten by a poisonous snake or suffered some other mishap, and that we had to head to the nearest hospital, a forty-minute drive away. If one of us was to step on a death adder, this would leave, on average, only five minutes for us to get to the car and start the drive to the hospital to receive the antivenom. This was not an idle consideration: on the previous trip, one of the assistants almost sat on a death adder that was cunningly resting between the fly and the entranceway to her tent. You learn to walk carefully and keep your eyes peeled in the Australian outback, though both these things are much more difficult in the predawn darkness.

Our objective was to search the ends of dead twigs for nice round holes made by twig-burrowing beetles. We were looking for twigs that had been vacated by the beetles and occupied by bees. Since we could not tell this in the dark, we plugged the ends of the twigs with cotton wool, marked them with flagging tape and moved on, looking for more holes in more twigs. Later, when the sun had been up for an hour or so, it would be warm enough for some of the bees in these nests to start flying. It was at this time that we went back to the twigs we'd marked earlier, plucked them from their bushes and stored them in an esky (Australian for beer—or bee—cooler) to be taken back to the laboratory at the end of the trip. On a good day, we found one nest per hour of searching. That's not a lot of positive reinforcement even on a day with a more leisurely start time.

But the inconveniences of extreme heat and extremely early starts faded into the background when compared to the fascinating research we were there to accomplish. The particular species of bee we were looking for, the trident-tailed exoneurella, has the largest-known size differences between queens and workers for the entire group of species it belongs to. The queens are odd-looking beasts indeed. With three prongs of approximately equal length at the end of a strangely flattened abdomen, they look something like ear-wigs, especially given that their wings are short in comparison with their body length. The workers are slightly more ordinary, with normal-size wings and a middle prong at the end of the abdomen longer than the other two. These bees use the flattened and toothy abdomen to plug the nest entrance to keep unwanted intruders out. And there were many potential intruders around—ants to steal the nest's contents, cuckoo bees whose offspring would eat the host's food, wasps or other bees that might want to take over, perhaps even individuals of the same species wanting to usurp the inhabitants' hard-won home. And, of course, there were intrepid melittologists trying to remove entire nests.

The trident-tailed exoneurella is an odd bee. It looks odd, its nest site is unusual, the way the adults care for the juveniles is curious and it has an interesting social life. But we can learn several important details about the basic lifecycle of bees from this unusual one.

Because a bee spends its entire juvenile phase in a nest with all its food provided for it, its life is strongly influenced by where its mother chooses to locate the nest and what she chooses to build it out of. This is where we will begin our story of the lives of bees.

~

You do not have to travel to the Australian outback to find bees nesting in dried stems or tree branches. You can find nests even in the depths of winter in cold temperate climates. This is most easily done by snapping the brittle stems of raspberry or black-berry bushes, or even old fennel stalks (though please do this with restraint, as opening a nest in winter will certainly result in the death of all the inhabitants). You will have more success if the hole at the end of the stem is nicely rounded, which suggests a well-maintained entrance. Small carpenter bees, usually dark greenish-blue with ivory-coloured markings on the face and legs, overwinter as adults in these stems and disperse in early summer—the males to search for mating opportunities, the females to start a new nest in a different stem.

Large carpenter bees do not look at all like their smaller close evolutionary relatives. Some large Virginia carpenter bees nest in the wood of an old grape arbour near the end of my urban garden, where my wife and I sometimes eat breakfast. The loud buzzing while the bees explore the wood can be quite entertaining. Several dime-size, perfectly round holes mark the entrances to their nests, which extend deep inside the structure. Sometimes I can hear them chewing. These bees have the odd habit of occasionally dropping out of the nest. Fortunately, so far, I have not had one land in my breakfast cereal or morning coffee. Fortunately, so far, they have not weakened the entire structure of the arbour to such an extent that it collapses.

Opening up the nests of these large bees is quite an experience. One of the highlights of the International Bee Course in Portal, Arizona, each year is Steve Buchmann's release of the carpenter bees. Steve is a pollination biologist renowned for his efforts to conserve pollinators and for co-authoring the first book to bring their plight to the public. Each year he spends a few hours in the Arizona desert looking for the dime-size holes in the dead flower stalks of century and agave plants that are the tell-tale signs of a carpenter bee nest. When he discovers one, he stuffs the hole with tissue paper and uses a machete to cut the stalk.

Back at the research station, the students and instructors gather in the shade of a giant cottonwood tree while Steve talks about the nests with half a dozen enormous bamboo-like poles stretched out before him. He picks one up and hacks at one end with his machete to crack the pole lengthwise. Immediately an extremely loud and angry buzz emanates from inside the nest; most students jump back a few paces (the instructors usually jump even farther back). As the crack in the pole grows longer, the bees inside begin to escape. The students yelp as one of the largest and blackest bees in North America flies towards them. But the bees have no thoughts of revenge—they just want to survive. (Unlike some people, bees are not vindictive creatures and they do not harbour grudges.) Nonetheless, it's a scary thing to have a large, angry-sounding black bee almost the size of a hummingbird fly straight towards you.

The developing larvae of these bees have a hollowed stem to grow up in. That doesn't sound very comfortable—and indeed, other bees provide a much more luxuriously appointed home for their offspring. Most leafcutter bees, as their name suggests, cut leaves. These bees are responsible for the almost perfectly semi-circular holes in the leaves of your prize rosebush. (Last summer, some of these bees became partial to the wild strawberry growing in my garden, and most of those leaves ended up with up to three semicircular cuts.) They line their brood cells with these leaf pieces, obtaining them with remarkable agility. The females sit on the edge of the leaf and start cutting with specially modified shearing mandibles. They rotate forwards as they slice, and do this in a way that at first seems daft. Like a lumberjack sitting on the tree branch he or she is cutting off, the leafcutter bee holds onto the piece of leaf she's collecting. She has to fly away with her prize at the exact moment her mandibles reach the end of the cut.

Members of the leafcutter bee family build their nests in the most diverse range of sites using the widest array of materials. Some line their nests with petals, plant hairs, resin, mud or gravel. Some nest in the ground, in holes in wood or even in abandoned snail shells. Resin bees, a specialized group of leafcutter bees, often have very long mandibles that function almost like forceps; they use these to hold a ball of resin, which they take back to their nests to make partitions between their brood cells or to plug the entrance with dense, sticky, impenetrable glue. The longest bee in the world (but perhaps not the largest in terms of overall mass) is the pluto resin bee, which was thought to be extinct until rediscovered on a small Indonesian island in 1981. It constructs its brood cells inside active arboreal termite nests. Doubtless the resin helps keep the termites away.

Leafcutter bees and their close relatives the orchard bees can choose nest sites that are inconvenient, and perhaps even lethal, for us. Orchard bees often line their brood cells with mud. This can be annoying if they choose to nest in the lock to your front door: the mud hardens almost to the consistency of concrete. It can be even more annoying if the species using your home security system as its base of operations is one that lines its brood cells with resin—while the mud may be prised out, resin is sticky and almost impossible to extract. In this instance you will need a new lock.

But things can be worse. During the First World War, a surgeon at a battlefield hospital found that bees were impeding his treatment of the wounded. Every time he hung up his stethoscope, a bee would start nesting and fill the earpieces with mud. Many times each day, the doctor had to extract the material with a penknife to permit him to hear the vital signs of his patients.

Bees were implicated in a more serious event when investigators at the site of a 1989 plane crash near Ruidoso, New Mexico, found some pollen and other plant material inside the fuel line of the downed aircraft. The families of the deceased claimed that a design malfunction had permitted the material to infiltrate the essential components of the plane, and they sued the manufacturers of the aircraft, the engine and the fuel control units. Melittologists were called in to investigate. A single bee hair was found in among the pollen, and it turned out to belong to one of the species that was reared from other parts of the crashed plane. The entomological evidence was unambiguous: the pollen ball had been constructed after the plane had crashed, not before. The bees could not have been the cause of the crash and were found not guilty.

In another instance, bees were the culprits but a person was blamed. During the Second World War, a series of aircraft crashes occurred at a military airfield in South Africa. The cause of the crashes was found to be a blockage of the airspeed indicator by fragments of vegetation. Sabotage was expected, and an air mechanic was charged with the crime. If he had been found guilty, he would likely have been executed. Luckily, an officer suggested that, though it was a one in a million chance, perhaps an insect had something to do with the blockages and recommended that S. H. Skaife, a well-known local entomologist, be called in. Under conditions of complete secrecy, Skaife was brought in to investigate, and he determined that the plant material resulted from the nesting activities of leafcutter bees. This time the bees were guilty, and the defendant was released.

Nesting in pre-existing cavities such as stethoscopes, the fuel lines of crashed aircraft, door locks or beetle burrows in dead twigs is not the most common choice for bees. Most bee species nest in the ground. Different species choose different types of soil; sandy soil is perhaps the most preferred, but clay and chalky soils are used by some bees. Others prefer to nest in dense lawns or patches of moss, some beneath a stone or the basal rosette of a dandelion. Some bees prefer a horizontal surface and others a vertical bank.

The depth of the nests they dig varies depending on the species and the details of the ground. In cold climates with short summers, nests are often shallow so the developing brood can benefit from the warmth of the sun's rays.

In the spring of 1986, I was surprised to find large numbers of the golden augochlorella in a roadside cutting on Cape Breton Island. This is a member of a tropical group of bees, most diverse in Brazil, with only a few species as far north as Canada. This was an almost Arctic outpost for an ancestrally tropical insect. At this study site, the females made their nests between large stones that were exposed at the surface of the embankment. The nests were very shallow, just a few centimetres below the surface of the soil. Using a simple laboratory thermometer, I demonstrated that warmth radiated out from the rocks long after the sun had set behind the nearby hills. The bees and their young were kept nice and warm late into the evening as a result of the mother's choice of nest site. But unfortunately, the nest site proved too hot for the bees in the year I studied them. It had been an unusually hot and dry year in Cape Breton, and the thin sandwiches of soil between rocks often heated up so much that some of the poor bee larvae were cooked to death. Those whose mothers had chosen a nest site with a rock on only one side generally survived, as did those that were not adjacent to any rocks. But those sandwiched between two sources of radiant heat were toasted, as indeed were some of the mothers that chose this nest site in the first place. This was a rare instance of large numbers of bees making the wrong choice because of unusual weather conditions in one particular year. Similarly lethal errors occur when bees nest in a riverbank that subsequently erodes or a sand dune that gets blown away in an unusually strong storm.

Bees that nest in the ground have to avoid getting cooked, they have to avoid getting frozen, they have to avoid getting waterlogged and they have to avoid getting desiccated. Not surprisingly, bees are often very choosy about precisely where they dig their nests, and they may spend several days trying out different pieces of potential real estate, occasionally burrowing down a centimetre or two in a test dig. Eventually, a spot is chosen and a serious burrow is excavated. The bee has to choose the right depth for the soil conditions. If the area will dry out substantially in summer, then the nest should be deep enough that there is sufficient moisture in the soil. If the offspring overwinter inside their brood cells, the mother has to dig down beneath the frost line.

Some nests are extremely deep, so to study their architecture, the melittologist has to dig down at least as deeply as the bees have. Several of my colleagues have reported digging up nests so deep that to get out, they had to carve steps in the side of the enormous hole their intellectual enquiry had caused them to excavate. In one case, a single bee had dug a nest that was over two metres deep, and the bee that did all this work was just one centimetre long. The hard-working nest digger moved at least five hundred times its own weight in soil. For me, that would be equivalent to moving fifty tons of dirt. If that sounds difficult, imagine having to carry that amount of soil, armful by armful, up a vertical tunnel.

The deepest nest I know of was excavated by Ronaldo Zucchi, a researcher at the University of São Paulo at Ribeirão Preto, Brazil. This nest had been dug by numerous bees; there were 884 golden-haired exomalopsis adults living in a nest that was over five metres deep, and it had probably been occupied for several generations.

～

There are numerous words that apply to bees because of the food they make for themselves (which we sometimes steal). Many of these—mellifluous, melodious, melittologist, etc.—are derived from the Greek word for honey, *meli*. But only those gastronomic extremists who like such things as the stinkiest of cheeses or delicacies such as *hákarl* (putrefied shark flesh that has mouldered, uncooked, in gravel for six months in Iceland) would appreciate the results of the labours of the vulture bees.

There are three known species of vulture bees—stingless bees that have switched from pollen to decomposing meat as a source of protein. They feed on this as adults and then provide the products of their digestion to the developing brood. Charles Michener, the dean of melittology at the University of Kansas, continues to publish research on bees, even though he recently celebrated his ninety-first birthday. In 2000, he published a revision of his Ph.D. thesis, which was originally published in 1944. (Another revision was published in 2007 and is almost a thousand pages long.) Mich, as he is called by friends, colleagues and students, describes the "honey pots" of vulture bees as containing "honey-like material or a paste made from honey-like material, masticated carrion, and bacteria." Opening a nest of one of these bees would be a malodorous, rather than mellifluous, experience. Sometimes these bees are predators, actively killing instead of waiting patiently for another organism to die. When a wasp nest is abandoned, for example, the defenceless larvae will often be eaten by vulture bees.

Less un-beelike are some cellophane bees, which provide a very liquid nectar mixture for their offspring. This fermenting soup is a breeding ground for yeasts, which provide most of the amino acids required by the developing larvae. The larvae float on a germ-ridden mead that is their sole source of food. The brood cells are waterproof and fitted with an undulating neck that may function like a fermentation lock. Digging up the nests of these bees, which are solitary but have very large brood cells, exposes the melittologist to odours reminiscent of some long-past hangover.

Some bees may have even more fluid provisions. The entomologist Hans Bänziger has studied a variety of unusual feeding patterns in tropical insects, such as blood-sucking vampire moths. He and his Thai colleagues recently suggested that there are several species of tropical Asian stingless bees that collect tears. Their published research article is decorated with numerous colour images of Bänziger's eyeballs being licked by little lachrymophagous bees. (These bees were also observed collecting eye secretions from several other mammal species.) The researchers suggest that the bees obtain most of the protein needed by their larvae from the antibacterial proteins in tears, and that an unusually distensible abdomen permits them to carry as much liquid as possible.

Some bees that collect pollen for protein don't collect nectar for carbohydrates but instead use floral oils. At least some of these species use the same oils to waterproof their brood cells. The pollen-collecting apparatus of these species has been modified into a dense, sponge-like brush that soaks up the oil. They can squeeze out the liquid using especially elongate spurs on their hind legs. There has been an interesting co-evolution between some of these oil bees and their floral hosts. The flowers these bees visit have oil at the end of their floral spurs. Most flowers with spurs have just one, which contains nectar, and the bees' tongues are long to access it. But the appropriately named twinspur flowers have a pair of floral spurs that house oil rather than nectar, and the bees that collect oil from these paired spurs use paired structures—their front legs—to do so. The co-evolutionary component here relates to the length of the spurs and the length of the forelegs of the bees. There now exists a complete sequence, from bees with normal front legs that visit flowers with very short spurs all the way through to bees with forelegs longer than their bodies that visit flowers with spurs longer than the width of the flower.

I would like to make one last distinction among bees based on their diet. Some bees collect pollen from a wide variety of different host flower species, while others are far narrower in their tastes and will collect pollen from only a single species, or a few closely related species. (Most bees will collect a wide variety of nectars even if they are very choosy about their pollen source.) There is a continuum from entirely promiscuous pollen foragers—generalists such as honey bees—to those extreme specialists that are entirely reliant upon a single flower species.

Consider the pollinators of rock nettle, a plant that grows in the deserts around Las Vegas. This is a beautiful silky, creamy yellow flower with a dense bunch of anthers in the middle. It is called rock nettle because it grows on rocky substrates, even vertical cliffs, and its leaves have stinging hairs. (How it manages to get enough moisture growing on vertical rock faces in the Nevada desert is a mystery to me, but clearly it can do so, at least in some years.) Various bee species visit this plant, and at least three of them have dark brown forequarters and a reddish abdomen (perhaps they're mimicking one another like the bees we discussed at the end of Chapter 4). These three species are from different families: melittid, halictid and andrenid. The first two are common, but the third is rare; it comes out only when it can be assured of a good quantity of its only floral host, whereas the other two species also collect pollen from close relatives of the rock nettle. The two less fussy eaters will appear when there has been enough winter rain to permit some of these relatives of rock nettle—plants that grow on flatter, and presumably somewhat less dry, substrates—to flower. But the fussiest bee, the andrenid, is an extreme specialist and will come out only after a winter of sufficient rainfall to promote profuse flowering of its single species of floral host.

Why do some bees specialize on pollen from only one or a few related plants? This topic has recently been investigated by the energetic young Swiss biologist Christophe Praz. We had a party for him when he visited my laboratory. At 11 p.m., after five hours of merry-making, everyone was ready to turn in for the night—except Christophe, who had decided that he wanted to keep partying, and indeed stayed out until well past one. But that was a well-deserved night off from his painstaking research. One of his most detailed studies involved gathering together the nests of four different specialist bees with different preferred hosts (one preferred asters, one brassicas, one buttercups and one viper's bugloss). He then carefully opened the nests and swapped the recently laid eggs from one species to another. To ensure a controlled experiment, he also transferred eggs to a different pollen ball made from the same species of pollen. He was clearly good at moving bee eggs around: well over 80 percent of the tiny, fragile bee embryos that he moved to the brood cell survived. The results were not so good for the eggs that were transferred to nests with different foods. Some bee species had almost a 100-percent death rate when given a novel pollen source. Two groups of plants, the asters and the buttercups, caused the death of all bee grubs that under natural circumstances would have fed on a different type of pollen. Buttercups are known to have toxic pollen, so the bees that specialize on these flowers must have evolved the capacity to detoxify the nasty chemicals in their diet. Asters didn't poison the bee larvae, but they caused trouble for the bees that didn't normally feed on their pollen because they are nutritionally deficient. Dandelion pollen is particularly difficult for bee larvae to digest because the wall of the pollen grain is hard to break down. (Imagine trying to live on a diet of Brazil nuts without the aid of a nutcracker.) Other asters are deficient in amino acids that the bees usually cannot get from any other source; presumably the bees that actually prefer aster pollen have another way of obtaining the necessary compounds. So this is at least a partial explanation of why some bees narrow their range of floral hosts. But how is it that the adult bees know which pollen to collect?

Christophe attempted to answer this question with another elegant set of experiments. He already knew that some specialist bees could develop from egg to adult when given nothing but the "wrong" pollen for food. Would bees reared on the "wrong" pollen collect this novel food for their own offspring, or would they revert to the pollen their species normally prefers, even though they had never experienced these flowers themselves? In other words, do these specialist bees imprint on the food they were raised on, or are they genetically programmed to collect the pollen of a specific plant? The answer was that these bees always preferred their species-specific diet. Not only did the females collect pollen only from their preferred flowers, completely ignoring the plant whose pollen they'd been raised on, but the males preferred to fly over these flowers when looking for mates, even if they had been raised on a diet obtained from a different plant species.

Of course, there is a middle ground between the extremes of using a single plant species for food and visiting almost anything with pollen. Some bees will concentrate on a few related plants in one locality but visit a different set of plants in another. Perhaps an extreme example of this is the band-footed sweat bee (a dull brown bee that produces one brood of workers each summer). In southern Ontario, workers of this bee forage almost entirely on staghorn sumac, but in Cape Breton, Nova Scotia, they forage on wild rose blossoms. Locally they are extreme specialists, but they couldn't choose two more different flowers to specialize on.

Some other bees like pollen from several completely different flowers. The red osmia, a common European bee with long orange-red hairs and a pair of horns on the face, is particularly fond of pollen from oak trees and buttercup flowers. Again, two more different floral hosts would be difficult to choose.

In general, though, we do not have enough information on the dietary preferences of bees. Too few have been studied in enough detail in enough places for us to be certain what their floral choices are. Some species were thought to be specialists until studied in detail, when it was found that they were actually collecting pollen from many different hosts. To be really certain of the diversity in the pollen-collecting preferences of bees, it is important to identify the source of the pollen grains that bees take back to their nests. This involves using identification guides to pollen grains (although the technology for obtaining DNA barcodes from pollen has been developed, it is currently very expensive). It does seem clear that the drier the habitat, the higher the proportion of bees that collect pollen from very few species of floral hosts. It has been suggested that up to 60 percent of bee species in deserts are very fussy pollen eaters. It was this that prompted my interest in the bees of the Atacama Desert, where this book began.

But whatever flower a bee collects pollen and nectar from, it collects a lot—enough for the complete development of its offspring. Imagine you, as a potential parent, building a room for your child-to-be and stocking it with eighteen years' worth of food from the supermarket. You give birth and put your newborn in the room—which is fairly secure against the elements and well hidden from natural enemies—closing the door behind you and leaving your darling to fend for itself. You can build another room for the next child without worrying anymore about the first. This sequence is repeated until you have enough offspring. No terrible twos, no teenage angst—your only responsibilities as a parent are to make the room, shop and produce the offspring. It is a "no muss, no fuss" approach to parenthood. This is the solitary bee way of mothering.

Still, collecting all that food for each offspring sounds like a lot of work. Eighteen years' worth of food would mean numerous trips to the supermarket, even if you had a half-ton truck. Exactly how much work this is for a bee has been summarized by Jack Neff, the director of the Central Texas Melittological Institute in Austin, Texas, and the most adept bee collector I have ever had the pleasure of working with in the field. He recently compiled statistics on all sorts of measurements related to the workloads of bees as they go about providing for their offspring.

Jack calculated that the total amount of food a female bee has to collect to provide for a single offspring varies from three to eight times her own weight. I am not sure how this compares to the human situation, but I would guess that both my children required more than eight times my weight in food between birth and adulthood. (It sometimes seemed that they needed that much each week.) Bees whose fully grown larvae construct a cocoon have to collect a larger amount of food for the simple reason that a substantial proportion gets turned into material for the cocoon. You too would have to eat more if you had to spit out a house to live in.

How many foraging trips are required to construct the pollen balls the larvae will feed on? The amount a female bee can carry varies considerably among species: some bees carry only one-fifteenth of their body weight per trip, while others can carry almost two-thirds of their weight at a time. Most bees probably make five to twenty foraging trips before they've collected enough food for the complete development of a daughter. (Sons usually require less food because males are smaller than females in most, but not all, bee species.)

How quickly a female completes a foraging trip is another highly variable factor. I have personally observed bees taking two hours to complete a foraging trip. (The population did not persist at that location because the females simply couldn't collect enough food to produce enough offspring.) The most efficient bees (and those closest to their food sources) can complete an entire foraging trip in a mere two minutes. The average trip for the nineteen species that Jack compiled data for took approximately fifteen minutes. That's much less time than most people spend inside a grocery store each week.

Females of the species that spend two minutes per trip can bring back thirteen times their own body weight in food per hour. Not surprisingly, these bees do not forage for long each day. In contrast, the least frenetically active females can amass only one-quarter of their body weight per hour.

The actual currency of importance to a bee is the number of offspring that can be provided for. Life is unpredictable, so the quicker a bee can reproduce the better. The maximum recorded daily rate for pollen-ball construction for a solitary bee is an astonishing nine—that is, enough food for nine offspring in a single day. That day's work required the mother to go on over 270 foraging trips. Even the most ardent "shop until you drop" advocate couldn't manage that in twenty-four hours. I must admit, though, that this extremely high rate of productivity has been found only under ideal greenhouse conditions. (Even if a person lived above a supermarket, it would be difficult to go shopping 270 times in one day—that's more than 11 times an hour for twenty-four hours, or once every five minutes for an eight-hour workday.) The same species managed only one or two pollen masses a day in an unmanaged setting. This would seem to be close to the maximum for most solitary bees under natural conditions: some provision only one brood cell every four days, and most bees take a day off after a stretch of heavy foraging work.

~

The lifecycle of a bee involves four stages. First is oviposition— the egg is laid. Second, the larva hatches and feeds. It rests safe in the confines of its brood cell with enough food to be able to complete its growth just by lying there and eating. When fully grown, the larva pupates. While it's in the pupal stage, its internal tissues are reorganized from larval to adult form, a metamorphosis akin to the transformation of a caterpillar into a butterfly. Last, it emerges as an adult bee.

The laying of an egg is not generally a remarkable act. But the size of the egg is worthy of note. Most bee eggs are banana-shaped and generally proportionate to the size of the female. A bee that is one centimetre long will generally lay an egg that is approximately two millimetres in length. But cuckoo bees lay disproportionately small eggs (they have to lay as many as they can once they gain entrance to a host nest, and smaller eggs allow them to have more at the ready), and carpenter bees lay disproportionately large ones. The largest bee egg measured was 16.5 millimetres long and was laid by a large carpenter bee. The longest known egg was twice as wide as the bee that laid it. That would be like a chicken laying an egg that was over thirty centimetres long.

Because the egg is laid on top of the pollen ball, the newly hatched larva can start eating even before it has crawled out of the eggshell (a bit like having breakfast in bed). Most bee larvae are fairly nondescript, legless white grubs. But the larvae of some bees that live on liquid pollen-and-nectar mixtures are exceptions. Some of these have rows of fleshy, paddle-shaped protuberances that let them float on top of their soupy food so they do not sink and drown.

In general, life for bee larvae is easy; they are safely housed inside a mostly waterproof brood cell and do nothing other than eat all day. But even though they eat all day, they defecate just once in their entire juvenile life. No diapers, no toilet training, just one defecation between "birth" and adulthood. Even if your child weren't in a sealed room all by herself this would seem advantageous in comparison with the human way of doing things. Why would bees have evolved to be like this? After all, they wouldn't actually have to change diapers even if their toddlers were wandering around their home.

Bees and their wasp relatives that feed on animal protein evolved from parasitoid wasps, most of which used their ovipositor to lay eggs inside another animal. The parasitoid larva eats away at the insides of its host until fully grown. Clearly, it's not advantageous to release waste matter into the confined environment of the animal whose insides you are consuming. (For a similar reason, most people would not locate their washroom inside the kitchen.) Even after the evolution of stinging and a pollen-eating diet, bees did not want to foul their brood cells. To avoid getting the rest of their food messy, they evolved to tighten the anal sphincter and hold all that nasty stuff inside until the last moment. This sounds rather difficult, but bees and their close relatives do not have any trouble controlling their processes of evacuation: their entomological equivalents of intestines and rectum are not even connected until late in their juvenile lives. They could not defecate earlier even if they wanted to. Once their intestines and rectum are connected, they let out all their waste in one extended dump. Then, after a day or more of continuous defecation, they are ready to pupate. (Okay, this may not be a crucial element of the economic importance and conservation-worthiness of bees, but it is one of those weird facts that makes life more interesting, and it is also an excellent factoid to bring up at a cocktail party.)

Of course, there are exceptions to these simple feeding regimes. For example, bumble bee larvae are clumped in small groups and are given food as they grow rather than all at once. This is called progressive provisioning, and it gives more opportunities for social interaction between adults and young (but still the offspring defecate only once).

Other bees that are progressively fed include the honey bees and the allodapine bees, such as the ones we were studying in the Australian outback at the beginning of this chapter. The allodapine bees are unique in that their larvae are structurally highly diverse, rather than morphologically monotonous like other bee larvae. Allodapines nest in hollowed stems, often with multiple adult females per nest. These bees have a crèche (nursery) at the end of the nesting tunnel where eggs are laid and larvae develop. The larvae have to move around in this crèche to get their food. As a result, allodapine larvae have all sorts of odd fleshy protuberances—humps and bristles to help them move around. They are remarkably diverse in appearance in comparison with other bee grubs, which sit immobile and eat away in the dark in their solitary brood cells like mollycoddled couch potatoes.

The fully grown larva, once having defecated, then pupates. Larvae of some bee species construct a cocoon before they pupate. Sometimes you can find these cocoons exposed in an earthen bank that has eroded since the bees vacated their natal home. Bee pupae are very fragile things—any slight wound is likely to result in infection, necrosis and death. But because they are safe within a brood cell, the only time most are likely to get wounded is in the first attempt at complete consumption by a predator. A cocoon gives the bee pupae even more protection against enemies and elements.

Emerging from their thin pupal casing, an adult bee has to harden its exoskeleton and pump blood into its wings so they expand to normal size. These "callow" adults then have to wait until their wings have hardened before they can fly. This defenceless part of their lifecycle takes place inside a brood cell that is hidden away from the bad things that can happen. It usually takes a couple of days before the new adult has become tough enough to leave its childhood home and take to the wing.

Most bees are solitary creatures. But those that live as adults in a nest with others are a different story altogether. Let's look now at their remarkably diverse ways of being social.

7

The Sociable Bee

DIGGING NESTS AFTER DARK IN CALGARY, ALBERTA

It was six in the morning and time to start digging up nests of the small shiny sweat bee. Because most bees nest in the ground, you have to dig them up to study them. This can be inconvenient. Because most bees are inactive until nine in the morning, studying them is usually a reasonably civil activity (in comparison with studying birds, say, when you really should be out there with binoculars ahead of the dawn chorus). But digging the nests of social bees requires a different approach. You want all the occupants to be at home to discover how many individuals there are in each nest, and to find the queen and the workers, as well as their brood.

In 1987 I was a postdoctoral researcher at the University of Calgary, studying with the bee geneticist Robin Owen. During regular working hours I was in the laboratory developing methods for genetic studies of bees, but before nine a.m. and after six p.m. I was out on the university farm digging up bee nests. This made for long days, especially in late June, the time of year most crucial to gathering data. Calgary is quite far north, and in June, towards the summer solstice, it stays light until after ten o'clock.

I started excavating nest 253 at eight o'clock one evening in late June. All nests to this point had been small. This was the time of year when broods in most nests had been produced by the queen acting alone, and as a result most had fewer than half a dozen brood cells and a simple structure, with the brood cells arising from the side of a short burrow. The whole nest was usually less than twenty centimetres deep, and finding the queen was not too difficult. Each nest took fifteen minutes to half an hour to excavate.

Nest 253 was different. The burrow branched just beneath the soil surface, and then branched again. There were dozens of brood cells along each sub-branch. How could I tell that I hadn't accidentally dug into several nests at once? I had blown talcum powder down the nest entrance. This also helped me find the adult bees: they were usually cowering at the end of the burrow covered in white powder. But in this nest, there were numerous branches for them to hide in.

The small shiny sweat bee is a social bee that normally has two to five workers per nest, and these workers are usually the offspring of the queen. Sometimes, overwintered females of this species stay in their natal nest and form a small society made up entirely of sisters. Nest 253 was an example of just such a sorority, and it had six adults in it on the day I excavated it, more than the average colony size in midsummer. These bees had worked to make the nest unusually extensive and unusually deep, which was both unexpected and somewhat annoying. I had started digging at eight o'clock; by ten, it was getting dark and I was nowhere near finished. It would have been hard to trace these narrow burrows (they're just two millimetres wide) through the ground even if they weren't already in a fairly deep hole. Fortunately, I had a flashlight with me. For the first—and hopefully last—time in my life I found myself digging a nest in the dark by flashlight. Nest 253 kept me digging until 11:20 p.m. At the end of each branch of the nest, I had to dig down many centimetres to ensure that there were no talc-covered females hiding from the exhausted, filthy melittologist.

Most bees are solitary. The female mates, constructs a nest, collects pollen and nectar to fashion into a pollen ball, lays an egg and closes the brood cell. She repeats this sequence until she dies, usually before her oldest offspring emerges as an adult. Even if there are millions of nests along the same riverbank, each individual female living there is solitary; she does not share her nest with others and will fiercely protect her home from any potential usurpers.

Other bees live in social groups of various forms and sizes. The simplest variant on the solitary lifestyle is a *subsocial* society in which the adult female interacts with her developing larvae by feeding them individually (the progressive provisioning mentioned in the previous chapter). Another comparatively simple form of social life for bees is a *communal* society in which they share a nest entrance and the first few centimetres of the burrow but otherwise act as if they were solitary. Each female constructs her own branch from the main burrow, builds her own brood cells, collects her own pollen and nectar, makes her own pollen balls and lays her own eggs. These communal bees are like people in a condominium: they use the same entrance but lead independent lives in their own apartments. What are the advantages of this egalitarian setup?

Research done almost thirty years ago at Cornell University in New York compared the rates of successful attack by the articulated nomad bee, a cuckoo bee, on the nests of the bicoloured agapostemon, a communal bee, and a closely related solitary species, the hoary agapostemon. Five of the six hoary agapostemon nests were successfully attacked by the nomad bee, and twenty-five of the thirty-four brood cells subsequently produced a cuckoo bee rather than a host. Successful attacks on the nests of the bicoloured agapostemon were much rarer. Indeed, only two solitary nests of this normally communal species were successfully attacked. The researchers concluded that the nomad bee had difficulty gaining access to the communal nests because there was always a female at the entrance. As soon as one bee left to collect food for her future offspring, another would take her place. The advantage these bees obtained by living together, even without complex social interactions, was the increased effectiveness of nest defence. If we all left our front doors open while we were at work or out shopping, I am sure living in a condominium would confer a similar advantage—there might be enough activity at the entranceway to dissuade thieves from their larcenous intent.

Other types of social grouping in bees involve a reproductive division of labour, usually between one queen and some workers. The queen lays all (or most) of the eggs and does not leave the nest to forage if she has workers available to go shopping for her. In contrast, the workers usually lay no (or few) eggs, do all the risky foraging and may also act as nest guards (though sometimes the queen will do this while all her minions are off foraging), attempting to repel ants, cuckoo bees and other potential enemies (including encroaching bee biologists). In other words, the queens reproduce, and the workers work.

A queen may be of the same generation, and hence approximately the same age, as her workers, or she may be their mother. Societies with a reproductive division of labour among females of the same generation are termed *semisocial*. Nest 253 housed this kind of society. Colonies where the mother is dominant are referred to as *eusocial*, which means "truly" social. Most eusocial colonies last for less than a year (bumble bees and almost all social sweat bees have such annual eusocial colonies). These colonies start with the nest foundresses in spring. These females are the sole survivors from the year before; they will have mated the previous fall and stored the sperm until they need it in the spring. Initially, the potential queen forages to produce a brood of workers, but after her first brood matures, she leaves the risky business of foraging to her daughters. In cold temperate regions such as southern Canada, most eusocial sweat bees produce only one brood of workers; these enlarge the nest and collect pollen and nectar to feed males and young queens. In warm temperate regions, such as parts of southern North America, the activities of the first brood of workers yield more workers, and it is only later in the year that males and next year's queens are produced. Populations of eusocial bees living in long-season environments can end up with multiple worker broods and hundreds of workers per nest. (The world record—held by a species with a nest that was excavated in Guatemala—is slightly more than one thousand.) Some bees have eusocial colonies that last for several years. The honey bee is the best-known example; stingless bee colonies also last for multiple years, as do the colonies of a few species of sweat bees.

Semisociality as the only type of social organization a colony exhibits seems to be rare. But many annual eusocial societies start off as semisocial ones in spring; the nest that kept me in a hole in the ground past my normal bedtime in Calgary in 1987 was an example of this. Semisocial societies are usually small, with between two and six females (one stays at home and lays the eggs while the others forage). It seems probable that most semisocial societies form when sisters overwinter within the same nest; they have a society all ready to go as soon as the weather warms up and flowers are available. But a semisocial society can also arise from a eusocial one if the queen dies. When a brood is orphaned, one of the workers becomes reproductively dominant. This often involves a great deal of conflict among nestmates, as all of the females "want" to be queen. The new "replacement" queen is usually either the largest or the oldest of the sisters in the nest. Similar royalty-replacing mechanisms seem to have operated in human societies, albeit in a more machismo manner, with the oldest son inheriting the throne but often having to do battle with other siblings to maintain dominance. In battle, the strongest one is usually the victor.

Eusocial societies are much more variable in size than semisocial ones. The golden augochlorella population that I studied in Cape Breton had an average of half a worker per nest. This meant that over half the nests were inhabited by a solitary female, while most of the rest had one worker and a few had two. At the other extreme are perennial eusocial bees, such as honey bees and stingless bees (like the vulture and tear-drinking bees), which can have tens of thousands of individuals per nest.

Many books have been written on the fascinating biology of honey bees, but what goes on in the colonies of the far more species-rich stingless bees has received considerably less attention. Here are some of the main ways in which the social lives of honey bees and stingless bees differ.

First, honey bees start a nest by swarming; the old queen leaves the old nest along with approximately half of the workforce while the new queen stays behind. Swarming is a rapid event. In the stingless bees, the workers gradually take nest-building materials from the old nest to the location of the new one. Only after several weeks or more does the new queen (not the old queen) leave the old nest to take up residence in her new royal quarters.

Second, honey bee larvae are fed progressively on a regular basis, whereas stingless bees lay eggs on a completed provision mass that will supply all the food required for the full development of the adult bee.

Third, in honey bees egg-laying by workers is rare and generally occurs only when the queen dies. In contrast, stingless bee workers often lay eggs (although these are usually destined to be eaten by the queen as an additional food source).

Fourth, as we have seen, honey bees have a remarkable facility for communication and can even direct other foragers to the best flower patches. Most stingless bees direct nestmates to good food sources by leaving a trail of pheromones between the hive and the booty, though some can use only the odours of the food itself as a cue.

Finally, in honey bees, it is the quality of the food provided to the developing larva that determines whether it will become a queen or a worker—those destined to become queens receive large amounts of royal jelly in their early development. In stingless bees, caste differentiation is more variable. In most species, it is simply the amount of food provided that determines whether an individual will become royalty or not. But in some of the species, there is a genetic factor: most females are genetically incapable of becoming a queen, and others will become one only if they also receive special food. Young queens of these latter species have been seen entering the nests of others, suggesting that they may be able to become a queen in the nest of a different family.

～

A reproductive division of labour is a bit of a conundrum for evolutionary biologists to explain. Why should a female bee stay at home and raise siblings rather than start her own nest and produce offspring? What advantage is there in evolving sterility? The answer to this seems to involve the unusual sex-determining mechanism found in bees.

In most organisms the genetic material is found on chromosomes, and chromosomes come in pairs. We call these organisms *diploid*. During the production of sex cells (sperm and eggs), the number of chromosomes is halved: each sperm and each egg has only one member from each pair of chromosomes. We call these cells *haploid*. The act of fertilization gives the egg the full complement of paired chromosomes, with one of each pair coming from the mother and one from the father. In other words, the fertilized egg becomes diploid through the fusion of two haploid sex cells. This means that each parent shares exactly half its genes with each of its offspring. Similarly, full siblings (that is, those with the same mother and the same father) share half of their genes with each other. Of course, all members of a species have the same set of genes; the important thing here is that those genes are identical by descent from immediate relatives. Within families of most animals, there is a 50 percent chance that any mutation in a parent, mother or father, will be passed on to an offspring, and there is a 50 percent chance that one offspring will share this mutation with a sibling.

That is the way sex influences genetic relatedness in most organisms, but in bees (as well as wasps and ants) the situation is different. Healthy male bees come from unfertilized eggs. The whole male bee is haploid, and barring mutations, all his sperm are genetically identical. Female bees are diploid. As a result, bees are called *haplodiploid*. Female bees that have the same father share an identical paternal genome; the half of a female's DNA that she inherits from her father is absolutely identical to the paternal half of the DNA in her sister. Because sister bees also share the usual 50 percent of their maternal genome, this gives them an average relatedness of 75 percent. Put another way, a mutation in a male will be passed on to all his offspring, whereas a mutation in a female will be passed on to only half of her offspring. The net result is that bee sisters inherit 75 percent of their genes as identical copies (and they are sometimes called supersisters as a result). This is more than the 50 percent of their genes they would share with their own offspring. A female with a mutation has a 75 percent chance of sharing it with a sister but only a 50 percent chance of passing it on to her daughter.

What this mathematics suggests is that a female bee gets a 25 percent higher genetic return on her investment if she helps to raise a sister than she does if she raises her own daughter. The genetic rewards for constructing a brood cell and collecting pollen and nectar are 75 percent if the work is to produce a sister and only 50 percent for a daughter.

This makes it sound as if any bee would rather be a worker than a mother. But we haven't included the production of males in our accounting. Males are an essential part of the reproductive cycle, but they mess up the elegant mathematics. The same genetic calculus results in a worker bee sharing only one-quarter of her genes with a brother. Because a male has no father, he has none of the paternal genes found in a sister, but he does have the normal 50 percent share of maternal genes. This gives an average of 25 percent of genes shared between a female bee and a brother. She would share half her genes with a son. The 25 percent benefit from sister production is cancelled by a 25 percent loss from producing a brother. The poor worker bee would seem not to benefit at all if she raises a brood with equal numbers of brothers and sisters. But there are two ways she can influence these genealogical asymmetries to her benefit. One is to produce more sisters than brothers. With more investments at a 25 percent profit and fewer at a 25 percent loss, you will end up with a positive balance at the end of the colony cycle. Studies of sex ratios in colonies of eusocial bees suggest that worker bees do manage to sway the balance in favour of a higher return on their investment by increasing sister production. But how do they do this?

This topic has been explored by the French biologist Cecile Plateaux-Quénu and her husband, Luc Plateaux. The couple recently celebrated a golden anniversary of scientific collaboration, which started in the late 1950s, just before Cecile completed her groundbreaking Ph.D. studies on the social behaviour of sweat bees. They also have something of the social insect's division of labour, but in this case it is taxonomically based: Cecile studies bees, while Luc studies ants. Of course they help each other too, so they live in more of a communal society. Cecile has worked for decades keeping ground-nesting sweat bees in the couple's basement, where she can control day length and temperature just as easily as in a laboratory. Her painstaking experiments led her to discover that if there is lots of pollen in the brood cell, a female bee is encouraged to lay a fertilized female-producing egg, rather than an unfertilized male-producing one. Cecile made this discovery by experimentally adding pollen to brood cells while females of her research organism, the calceate sweat bee, were out foraging. She gave each brood cell a large boost in provisions at the end of the female bee's workday, while the mother was out collecting nectar. The returning females showed variation in how they responded to the novelty of finding additional food in the brood cell. Some of them just continued as normal, exhibiting no surprise or gratitude. Others ran about the nest in apparent confusion. How would you respond if you lived alone but returned home one day and found vast amounts of food that you knew you had not purchased? Some of Cecile's bees were so put off by the unexpected gift that they left their nest, apparently checking whether they had accidentally entered the wrong burrow. But eventually, all the females settled down and laid their eggs. These eggs produced disproportionately more daughters than they would have if the unsuspecting females had not had their food stores enlarged (that was the case in the control nests, which were opened but did not receive additional pollen). This suggests that a worker might be able to influence the sex of her mother's next offspring and obtain a higher genetic return on her investment simply by working harder.

The other way a female bee can benefit from the mathematics of genetic relatedness is by raising a combination of sisters and sons rather than sisters and brothers. This way, she gets all the benefits of the increased relatedness to sisters (75 percent) over daughters (50 percent) and none of the disadvantages of the decreased relatedness to brothers (25 percent) over sons (50 percent). This would appear to be a win-win situation. But the snag is that it applies to each individual worker in the colony, and each worker "wants" to lay her own male-destined eggs. A worker does not want to raise her sister's sons because she is less genetically related to her nephews (37.5 percent) than she is to her sons (50 percent). Similarly, the queen wants all the males produced in the colony to be her sons rather than her grandsons. In a eusocial bee colony, genetic analysis predicts that all the females will want only the queen to produce daughters, but all the females want to produce sons. This suggests that there is a considerable conflict of interest over the production of males. Not surprisingly, then, when it is the time of year to produce males, fights break out among competing female nestmates, all of whom want the colony to raise their own sons. Chaos erupts in what previously appeared to be a harmonious society. While this event has not been studied in detail in the sweat bees, it is well known that when a bumble bee colony switches to producing males, all hell breaks loose. At this switch point, workers fight one another, they fight and sometimes kill the queen, and they throw larvae out of their brood cells in fits of seemingly temperamental infanticide.

But all these calculations make sense only if there is one mother and one father in a monogamous family. If there is more than one father, then relatedness among sisters decreases below the 75 percent supersister value and can easily become less than the relatedness between a mother and her daughter. This suggests that potential queens might avoid some conflict in their homes by restricting their mating activities and having sex just once in their entire lives. For some bumble bees at least, monogamy seems to be the rule, and it seems to be the ancestral condition for all groups of Hymenoptera that have evolved queen and worker castes. Honey bee queens are much more promiscuous, mating with and storing the sperm from over a dozen different males. In a honey bee hive, the workers would be more closely related to their own offspring than they are to the (mostly) half-sisters that they work so hard to raise. But honey bees have reached a social point of no return: a single worker couldn't start a nest all alone even if she was able to mate and lay fertilized eggs. The entire ecology of honey bees relies upon the continuous flow of information among numerous nestmates, each with an assigned set of tasks. A single honey bee worker could no more easily rebel and start a colony alone than a single army ant could go on a swarm raid all by itself.

These relatedness arguments suggest that bees that live in colonies should be able to tell relatives from non-relatives, friend from foe. This has been investigated in a clever series of experiments performed over the years by students working under the supervision of Charles Michener at the University of Kansas. The zephyr sweat bee, a close relative of the species whose nests I was digging at the beginning of this chapter, is the laboratory rat of choice for bee social biology research in North America. Artificially reared colonies were kept for many generations at the University of Kansas. The researchers took advantage of what they knew about the social dynamics in orphaned colonies, where one bee would become queen, another a guard and the others foragers. By putting individuals from different colonies into the same artificial laboratory nest, the researchers could produce "families" of unrelated bees in something like a small orphanage. They could then see which other bees the guards would let into the nest. Interestingly, the researchers found that while the guards would fend off their own supersisters if they'd been reared in other nests, they would let in the supersisters of their nestmates. The guards even let in the nestmates' supersisters that had been reared in the same nests as their own supersisters (whose attempts at nest entry they repulsed). The explanation for this unexpected result is that the guard learns the smells of its nestmates. If a bee smells like a guard's nestmate, she will let it in; if it doesn't, she will repel it. Supersisters smell very similar to each other because the odours that bees produce to tell each other apart have a genetic basis. The supersisters of the guard do not smell like nestmates, even though they smell very similar to the guard herself. This demonstrates that the guards do not smell their own body odour and use that information to decide who they are related to.

These results sound counterintuitive until we realize what a dastardly set of circumstances the experimenters placed the guards in. Over the approximately twenty million years that these bees have resided in eusocial societies, they have generally lived in comparatively simple family units. Only in the 1970s, with the start of the research in Kansas, did the guards find themselves in a situation in which all their nestmates were genetically unrelated to them. Their responses make sense under the circumstances they would be "expecting" to experience. It's not their fault that they responded in a seemingly illogical way to an entirely novel situation.

~

Not all researchers believe that genetic genealogies are the best way to explain the origin of a reproductive division of labour in bees, and some genealogical analyses of social bee families suggest that workers do not always capitalize on the potential genetic benefits of sister-rearing. One of the counterarguments some researchers make is that the nest itself is such a marvellous resource that staying home to raise relatives is a better bet than running the risks of finding a new nest site and expending all that energy constructing a new burrow. But assessing the benefits of inheriting a home is more difficult than measuring genetic relatednesses among nestmates. The latter requires no more than some fancy genetics tools, but the former demands extremely long hours of fieldwork, with the researcher getting to know each bee as an individual. Critics of the relatedness approach have rightly pointed out that most researchers have taken the easy way out and estimated genetic genealogies rather than the ecological costs and benefits associated with being a worker.

Still, some researchers have argued that insect societies arise because of manipulation by the mother, who controls the size of her worker offspring simply by varying the amount of food she collects before laying the egg. She can also make sure that she does not produce more potentially rebellious offspring than she can dominate.

Despite the almost fifty years of research aimed at determining how insect societies evolve in the first place, no consensus has been reached. I believe the crucial studies remain to be done.

We have discovered that sex in bees seems to influence their social lives in unusual ways. In the next chapter, we will investigate how the same sex-determining mechanism that promotes social living also increases the risk of extinction.

8

Sex and Death in Bees

CHOOSING A MATE IN SUBTROPICAL FLORIDA

It was foggy at nine in the morning at Knight's Key Campground in Florida, and the bees were rather slow waking up. I paced back and forth among the beggar's ticks plants (the favoured food plant and sexual rendezvous point for the bee I was studying), looking for poey's gregarious bees that were attempting to mate. I had planned to spend all day looking, so I was grateful for the cooler weather and fog. I was staring at bees mating on flowers all day to find out how choosy the males were, and to determine what cues they used to tell the females apart.

The early literature on mating strategies often suggested that male animals—and I include humans in this—should mate with any female they can (a veritable bestiary of Don Juans). I have stated that this was wishful thinking by earlier generations of biologists, who were mostly male. The argument that males should mate at every possible opportunity makes many assumptions.

First, it suggests that sperm is available in infinite supply—or at least in such large quantities that a male can mate with an unlimited number of females. It seems unlikely that either of these possibilities is true, especially in bees. A male bee produces all his sperm while still a pupa. This sounds rather precocious, but he is fully potent almost as soon as he can fly. Male bees certainly have a finite supply of sperm.

Second, it assumes that the only contribution males make to their offspring's future is the sperm that initiates their development. This is the case for many animals, but it is far from true for others. (I doubt that fathers reading this would agree that they have provided nothing for their children since giving them a start in life at the zygote stage.) There are some male crickets that provide an enormous meal to the females they mate with. This nuptial meal is made up of glandular secretions and can be up to 40 percent of the male's weight: a considerable investment in his offspring—equivalent to a human weighing one hundred pounds producing a forty-pound baby. However, male bees do not provide much in the way of additional resources for their offspring, and they do not work in the nest (although there are some examples of male bees keeping their larval siblings warm by sitting on their brood cells).

Third, it assumes that a female will use sperm from any male she has mated with. But females can be very choosy about what sperm they will use, even rejecting males whose sperm they have already received. Furthermore, we have already seen with the social bees that many females won't use any sperm at all. The reproductive division of labour means that workers are unlikely to produce any offspring and therefore would be a bad choice for a male to mate with. The hypothesis that males would choose queen-like females over worker-like females was what I was investigating in Florida that foggy morning.

Poey's gregarious bee lives in societies that are predominantly eusocial, but it is unlike most eusocial sweat bees, which start their nests in spring, have workers that are active in summer and produce males mostly at the end of summer. Here in the balmy climes of the Florida Keys, this species produces new queens all year round, workers all year round and males all year round. For other social bees with an annual colony cycle, males are found mostly towards the end of the cycle, when the remaining workers are old and the young females are virgin queens. Telling these two groups of potential mates apart is not too difficult. But in the Florida Keys, male poey's gregarious bees are faced with old workers, young workers, young virgin queens and not-so-young queens that have already mated. And they are faced with all these different types of females all the time. For a male, all these females are potential mating partners, but each provides very different genetic rewards for his mating efforts. In the evolutionary currency of number of offspring produced, a sterile worker is an entirely unsuitable mate, whereas any male should want to be the consort of a virgin queen. As an added complexity, some workers of this species do lay quite a few eggs, and so mating with a young worker may not be as bad a strategy as it is for males of other species. Nonetheless, a male would best choose to be prince to a virgin princess that is soon to become queen rather than to dote over her older mother or a scullery maid of any age.

I watched the males in their mating dance. They had three different techniques—they would approach a female and then fly away; they would approach a female and make brief contact before flying away; or they would fly at a female at great speed, knocking her off the flower into the undergrowth below in a serious attempt at copulation. By collecting the females that elicited these different responses, I hoped to discover what cues the males were using to make their decisions on how to behave. (I tried to catch both the female and the male that responded to her, but the sheer rapidity of flight and the usually instant responses of the males made this almost impossible.)

The males should have made mating with young queens their top priority. These were the females that would soon be starting new colonies, and therefore were most likely to use lots of the male's sperm to fertilize eggs. They would be large and newly emerged. Young workers, especially larger ones, might also use some sperm if they were able to lay fertilized eggs, but older workers were unlikely to produce any offspring at all. It was easy for me to tell the old bees from the young ones because their wings were frayed and discoloured with age and they had shorter mandibles (these wear away as they excavate the nest burrows and brood cells). It remained to be demonstrated that the males could make similar assessments.

The choice for the males seemed clear—go for young females, and preferably large ones. Perhaps unsurprisingly, this is what my data suggested the males mostly did. The females that were knocked off the flowers into the surrounding vegetation in a bee-like version of a parsimonious courtship were larger and younger than those that the males just touched before flying away. The females the males decided not to even approach were generally very old; perhaps the males could detect the fraying or discolouration of the wings. (I don't think they were looking at the mandibles to distinguish among potential mating partners; they usually approached from the other end.) The males seemed to be behaving in the way an evolutionary biologist might expect them to, but there were exceptions. Not every female that was knocked off a flower was young or large.

Why might males make mistakes like this? I could only assess the characteristics I could see with a microscope. If these females were giving off different chemicals, or different chemical blends, I would not know about it. If an aging female worker was foraging on a flower with an attractive virgin queen nearby, the male could have been confused by the chemical cues into thinking that the female he was orienting towards was potentially fecund royalty rather than an ancient member of the labour force (not a bee that would normally have aroused his interest).

A male's lack of experience could also have contributed to the apparent errors. It could be that it takes trial and error for a young male, fresh out of the nest, to be able to tell virgin queens from other females. It may be necessary for a male to make contact with a few less suitable mates before learning the chemical or other cues that tell him which females are more likely to accept his advances and use his sperm. To meet a fecund virgin queen, you have to kiss a lot of old workers.

∾

Finding mates is a necessary part of the reproductive process for all organisms that have sexual reproduction. But the outdoor world is a complex place for a small insect, especially one that has perhaps a total of two weeks to live, has a very small brain and has to find virgin females before any competing males do. This requirement causes male bees to become adults slightly earlier than females do—a phenomenon called *protandry* (literally, "males first").

Not surprisingly, reproductively competent bees of both sexes have evolved strategies to ensure that mating can be done in an efficient and timely manner. Like young humans, they congregate at bars—but for bees, these are nectar bars rather than establishments that provide fermented products. A nice clump of flowers is a good place for a male bee to search for females. Virgin females have to feed on nectar to fuel their flight and on pollen to start growing the comparatively enormous eggs they lay, and they have to collect both to construct a pollen ball. Virgin females will visit flowers as soon as they are capable of flight. Males may even be able to entice an already mated female into mating while she's collecting food for her offspring.

Males of the wool carder bee have taken advantage of the females' penchant for shaving leaves as part of their strategy to obtain mates. Wool carder bees are as long as but quite a lot wider than honey bees, and they have lots of bright yellow markings on their bodies; they look like overweight, hirsute wasps. The females shave hairy leaves to make a soft, downy lining for the brood cells for their off-spring. If you listen carefully, you can actually hear their mandibles scraping away at the leaves of lamb's ears and other hairy plants. Because hairy leaves are a limited resource, males will defend patches of suitable plants against all comers; I often see them battling other insects around the sage flowers in my backyard. Upon detecting an interloper, the male wool carder bee approaches. If it is another wool carder bee, he inspects it to see if it has the facial markings of a female (less yellow) or a male. If the bee is a female, then mating, or at least an attempt at mating, results. If it's a male, then the territory-holder buzzes, head-butts and chases the intruder away. If it is another type of bee, he will bump into it in attempts to get it to move elsewhere. But honey bee workers are somewhat immune to such subtle persuasion: they will return repeatedly, despite being crashed into several times. Male wool carder bees have special adaptations to deal with such persistent territorial violations. They have four robust spines at the apex of their abdomen, and they use these to punch holes in the exoskeleton of bees that persistently invade their territory.

Nest sites are another good location for bees to meet and mate. Approaching females when they first emerge from the ground is a good way of getting to them before anyone else. Given that bees are generally protandrous, there will be many more males flying over the nest site than there are virgin females emerging from the ground below. As soon as a newly emerged female moves close to the soil surface, a whole bunch of males will crowd around. This can become quite extreme: a dozen or more males may form a tight, writhing ball around a single female. Some males, and even some females, can be killed in the crush. Males of Dawson's burrowing bee, a very large, ginger-haired species found in Western Australia, will dig down towards an emerging female and try to mate with her before she has even seen the light of day.

Nectar bars, nest sites and nest lining stores are not the only locations that bees use as rendezvous points for mating. Look at the hedgerow in your garden, especially if the leaves are shiny, on a warm, sunny spring day. On the sunny side of the hedge, you should see male bees cruising to and fro, looking for females that are attracted to the bright reflections. Sometimes the males will be so numerous they'll appear to swarm around the bush, and they always outnumber their potential mates (unless the rendezvous point also has floral resources for the females to collect for their offspring).

This may seem somewhat sexist, with the male bees being choosy about the females and burrowing down into the ground to find the youngest ones. But things are not so one-sided. Consider the males of some of the pearly-banded bees. These bees have extremely elaborate modifications on almost all parts of their bodies, including unusually shaped mandibles, spectacularly dense beards, antennae that look almost like corkscrews and legs that are swollen to make room for muscles that would make an Olympic weightlifter look like a skinny weakling. The undersides of their abdomens are replete with rows of spines like the teeth on a comb, and may also have long, curlicued flaps and dense brushes of hairs; their genitalia are as elaborate as those of any other bee (and male bees have genitalia that are often remarkably complicated). The males look like this because the females prefer them this way. They want males that can tickle them with their long beards, prod them with their unusual spines, comb them with their rows of bristles, hold them tight with their muscular legs and do all sorts of odd things to them with their tail ends. Males that can't do all these things—and do them at the same time or in a specific preferred order—are destined to be rejected as partners. Most male bees do not have such a large array of sexually selected idiosyncrasies, but all have some of them. The more extreme modifications often make the poor males look quite ungainly, and they must certainly hinder their ability to function in ways other than wooing a female. For example, some of these adaptations make it difficult for the male bee to keep his wings clean; others make it difficult for him to walk on the surface of a flower or to gain purchase on the burrow walls as he climbs out of his natal nest to begin seeking an end to his bachelorhood. The males have evolved such clumsy and otherwise disadvantageous morphologies as a result of sexual selection by the females.

These morphological modifications are not the only thing that determines the males' mating success; sometimes they have to smell just right. Female orchid bees make the males go to considerable lengths to obtain a complex odour before they will mate with them. Orchid bees are spectacularly beautiful—some are brilliant metallic green, some are cerulean blue, some are purple with bright red tips to their abdomen. They are called orchid bees for the obvious reason that they visit orchids. But it is the males that have the fondness for the flowers, from which they collect fragrances to charm female orchid bees. The males have extraordinarily expanded hind legs, but these are not burly for battle or grotesquely swollen for grappling with females during amorous embraces. No, their huge hind legs serve as perfume bottles. The insides of the enlarged hind legs contain sponge-like tissues rather than muscles, and their purpose is to store as many different fragrances as possible. The more complex the bouquet, the more popular the male. Orchids are pretty rare in the rainforest, although there are many species of them. To collect the desired array of perfumes that will make him a favourite with the females, a male has to work hard and visit a large number of orchid species. This gives the romantic advantage to the older males. One bizarre consequence is that younger males will raid the perfume storage organs of dead males (even recently deceased males smell better to a female orchid bee than young ones).

This desire for fragrances among male orchid bees is of assistance to researchers because we can put out odorous baits to attract them. You can get hundreds of male orchid bees at a single bait patch. The baits themselves are varied and include substances you may have smelled, such as vanillin, and others you may have wished you hadn't. Scatol falls in the latter category, and yes, it does refer to the scientific name for an animal dropping, a scat. If you are taking some of this chemical on a field trip to study orchid bees, you'd better hope the bottle doesn't break open in your luggage; it would be difficult explaining that smell to a customs officer.

Female orchid bees place great demands on the males, but females of all bee species can exert more control in the mating game than females of most other organisms because they can reproduce without having mated at all. Virgin birth is common in bees because all healthy males are produced either by a virgin or by a mated female that does not fertilize the egg as she lays it. This unusual sex-determining mechanism means that a female can produce sons even if she can't find a mate, but it also has the unfortunate side effect of producing a certain proportion of dud males, and this can increase the risk of extinction in bees to surprisingly high levels. To explain the role of sex in extinction risk in bees, we have to delve into how sex is determined.

~

In humans, mice, fruit flies and indeed most animals, sex is determined by chromosomes. If you have two X chromosomes—XX—you are a female, and if you have one X and one Y—XY—you are a male. But in bees, an individual's sex is determined by a single gene rather than a whole chromosome; this gene is called the *sex locus,* and it comes with many variants, called *sex alleles.*

The genetic trick here is that to produce a daughter, a bee has to fertilize her egg with a sperm that has a different sex allele. In other words, for a female to be produced, the fertilized egg (zygote) has to have two different sex alleles, one from Mom and a different one from Dad. (Technically, the fertilized egg has to be *heterozygous* at the sex locus for a female to be produced.) If the egg the female lays is unfertilized, then it has only one copy of the sex locus and can have only one type of sex allele, and the result is a healthy haploid son.

But there is another way for an egg to be laid with only one type of sex allele in its genetic makeup. If the sperm and the egg it fertilizes carry the same sex allele, then the egg will have two copies of the same sex locus variant. Technically, this individual is *homozygous* at the sex locus, and so is also a male. But males that have two copies of the same allele (rather than just one copy of all their genes) are duds: they either die before being able to mate, produce defective sperm or produce offspring with three sets of chromosomes.

By definition, a female bee has two different alleles—let's refer to them as Q and R—at the sex locus. She can produce Q and R sons without mating. If she mates with a male P, she can produce PQ and PR daughters, and everything will be fine. The male offspring will be haploid and the daughters will be diploid. But if she mates with a male Q, half her fertilized eggs will be QR and half will be QQ. The QR offspring will be healthy diploid females, but the QQ offspring will have only one version of the sex gene and will be dud diploid males.

What a daft way to do things. Clearly the production of diploid males is a drain on the resources available to a solitary female bee, a social colony and an entire population. Indeed, it is the largest genetic drain known, at least in theory, and therefore has the potential to be the largest genetic influence on extinction for any organism. This is why worker honey bees have evolved to recognize diploid male eggs and kill them as soon as possible, thereby not wasting resources on dead-end males. Solitary bees and mass provisioning social ones do not usually have the luxury of interacting with their developing juveniles and so cannot diminish the impact of diploid male production with such ease.

Normally, comparatively few diploid males are produced in any bee population because the sex-determining locus is hypervariable: there are many more alleles of the sex locus in a population than there are of other genes. It would be like having twenty different versions for human eye colour. Imagine what a rainbow of people that would produce. With twenty sex alleles in a population, only

5 percent of the diploid individuals are expected to be genetically dead-end diploid males. That's a comparatively minor drain on the population in comparison with what happens when there are only five sex alleles. In that case, 20 percent of the diploid individuals would be dud males.

What determines the number of sex alleles in a population? Primarily, it is the number of individuals: the larger the population, the more genetic variation it will harbour. A larger population will have more sex alleles and will produce a smaller proportion of dud males, while a smaller population will have fewer sex alleles and a larger proportion of dud males.

These dead-end males are expected to be a drain on bee populations and cause an increased risk of extinction that most other organisms do not face. But what is the magnitude of this? This is a difficult issue to study because the negative impacts are expected to take place over longer time frames than can be studied as part of a thesis project or a government-research grant. Biologists investigating long-term processes are in a difficult situation. How can we study phenomena that take place over decades or even centuries when we don't have confirmed long-term research funding and maybe won't even live that long? Computer simulations are a quick and painless way of understanding these processes.

I willingly admit to being rather inept with computers. I can turn them on, create documents and write emails, but that's about it. Fortunately, just when it became important to find out what long-term impact diploid males were having on bee populations, a student with the requisite skills, Amro Zayed, came to work in my laboratory. Amro's first research project had involved sampling bees along a transect in Florida; he then spent three weeks in the jungles of Panama, and as we will find out later, he'd also spent several months doing fieldwork in the driest desert in the world. After all that, he wanted to spend some time in the safety of his own apartment. And he did just that. He spent months at his desk writing computer programs to model the effects of the production of diploid males on the persistence of bee populations and comparing the results with similar simulations for organisms that don't have the unusual sex-determining mechanism found in bees.

The results were dramatic. Once their populations reach a small size, Amro proved, bees are far more likely to go extinct than organisms that have the normal chromosomal method of sex determination. Their probability of extinction was almost ten times higher than that in organisms already known to be at risk for different genetic reasons, and this was entirely due to their unusual method of sex determination.

One other conclusion that could be drawn from the simulations was that there will be increasing numbers of males in the final generations of a bee population that is on the way to extinction. There will be the expected healthy haploid males and some diploid females, but a growing proportion of the diploids will also be male. The last generations of a dying bee population will be heavily male-biased (a finding that is supported by observations of dwindling laboratory colonies of wasps with the same sex-determining mechanism).

Again, these results were a bit of a surprise. The dogma at the time was that in haplodiploids, such as bees, bad alleles would get flushed out of the population through the haploid males. Because healthy male bees are haploid, and because haploids have only one copy of every gene, they have to express whatever they have, even if it will make them unhealthy or kill them. The sex-determining mechanism of haplodiploids seems to be a way of keeping their genes free of undesirable negative mutations. The bad alleles have to be expressed by the males that inherit them, but these males don't do well, and so the less healthy genetic variants gradually get whittled away. If it's a new mutation that has a negative impact, it may never get to spread through the population in the first place.

But the problem with haplodiploidy as a means of sex determination is that it has both sex and death built into it. The only way to produce a healthy, fecund female is with two different alleles at the sex locus. But two copies of the same allele results in death, sterility (which for natural selection is just as bad) or the production of sterile triploid offspring (which is making things worse by extending the impact to the next generation). In this way, the sex alleles function like lethal alleles, which are genetic variants that cause death when they are expressed.

Amro's computer programs suggest that bee populations are particularly susceptible to extinction. Some of the most important organisms on the planet are at risk from the effects of small population sizes and will go extinct comparatively quickly.

This suggested to me that bees should be considered canaries in the ecological coal mine. If bee populations are in trouble, it would indicate that the ecosystem in which they live is not sustainable. Declines in bee populations will be an early warning signal that things are going wrong in the great outdoors. Studies of wild bee populations might therefore give us advanced information on the state of the world.

But how can we make use of the information that the thousands of species of bees might provide to us? A first step is to find out how to count them, a topic we will investigate in the next chapter.

9

Where the Bee Sucks, There Hunt I

PAINFUL BEE SAMPLING IN THE TEHUACAN DESERT, MEXICO

With one of the first swipes of my net, several fine cactus spines became lodged in my thumb—nasty things with hooks on the end that are almost impossible to pull out intact. I predicted (correctly, as it turned out) that the short, curved ends embedded in my flesh would cause an infection over the next few days. A few minutes later, a large acacia thorn pierced the sole of my shoe and stuck in my foot. Less than an hour later and another cactus spine ended up under a fingernail. Next, a fearsome pointed tip of an agave leaf pierced my left thigh to the depth of about an inch. But at least I was now almost symmetrical: my right thigh had been similarly punctured the day before, albeit higher up and less deeply. I was also stung five times. Three of those were on the thumb injured by the cactus spines, and it was throbbing badly.

More interesting than these fairly common desert bee-hunting mishaps was the small tree with hairy branches that I ran into. The hairs had urticating properties—in other words, they stung. It felt as if someone were holding a few burning candles against my left arm and rib cage. I pointed the offending shrub out to Eric Ramirez, a

Mexican student who was accompanying me in the field, and asked him what the tree was called.

"It is called bad woman," he said. When I complained that the term was sexist, he told me there was another pain-inducing shrub called "bad man," and he said the masculine-monikered version was even more painful than the feminine one. In comparison with most bee stings, with which I have considerable experience, the pain inflicted by this tree was nasty.

Late in the evening, I experienced another interesting and unpleasant situation—I had taken a few bites of a fruit called zapote negro. It looks like a large green persimmon, but when it is ripe the insides turn black and taste like treacle. The one I bit into was unripe. The effect was not unlike having my digestive tract washed in a strong extract of ginger and chili powder. I was burning inside and feeling extremely nauseous. I hoped that this was not one of those fruits that are nice when ripe but fatally poisonous when eaten before their time. I was barely suppressing the vomit reflex when I caught my last bee of the day, a large carpenter bee hovering around an acacia bush.

I endured these irritations just as I did the normal ones usually associated with my fieldwork—the thorny grass and seeds that stuck in my socks and scratched at my skin at every step, the chigger bites that itched so badly in areas where clothing was tight. Conditions that day were also more humid than I was used to in deserts, and somewhat unusually for a day in the field, I was damp from sweat. It was now the end of the day, and I was suffering from burning rashes, nausea, puncture wounds, stings and scratches. I was sticky and thirsty as hell. But I was very, very happy. It had been a great day. I had caught perhaps sixty species of bees, and this does not happen often.

I found the pains and irritations easy to deal with simply because I have such a passion for the work. Bees completely fascinate me, and I can put up with a considerable amount of discomfort while looking for them. Indeed, my concentration on the task is usually sufficiently intense that I hardly notice these irritations, except at the end of the day, when the bee sampling is over and I have to put up a tent in the middle of nowhere in the dark. It is this level of impassioned concentration, combined with perhaps an inadvisable lack of regard for personal comfort or even safety, that leads me to swing my net among cactus spines and wander after bees without noticing the enormous agave daggers aimed at my legs. Surveying bees in deserts is my passion, and on this day I was in heaven.

This trip was a comparatively luxurious one. I was spending a few days with Peter Kevan, a Canadian pollination biologist from the University of Guelph, discussing some research projects, sampling bees and working with a few students of a Mexican colleague, Carlos Vergara. We were studying bees and pollination in a botanical garden in the Tehuacan Desert in Puebla State, Mexico. Even though there was no electricity at the field station, there was gas for a lamp at night, as well as cooking equipment, chairs and a table, and a comfortable bed. For most of my field trips, I get by with a Bunsen burner to heat coffee in the morning and cook food in the evening. After a long day surveying bees, I usually have to process the samples while sitting in a rented half-ton truck with only the dome light to help me guide the pins into the bees I've collected.

Still, the rigours of my worst trips are nothing compared to the conditions experienced by the trailblazers of the past, those Victorians who travelled through uncharted territory before the invention of air travel, refrigeration and vaccines or medication for diarrhea. Travelling by donkey or even on foot, these folks would often need weeks to go as far as we can in a mere two hours. The inconvenience of a delayed flight is nothing compared to having to eat your mode of transport to avoid starvation. Sometimes the explorers perished. But those who returned with their samples left behind a legacy of numerous newly discovered species. My objective is similar—discovery. It is the excitement associated with this that permitted me to ignore the cactus and acacia spines, various stings and burns, and the heat.

Bees are actually almost everywhere—wherever there are flowers, there will usually be bees. From over four hundred metres below sea level on the shores of the Dead Sea, in Israel, to over forty-five hundred metres in altitude in some of the world's highest mountains, there are bees—and particularly interesting ones in each place too. Bees are found on every continent except Antarctica. They can be found on the edge of absolute deserts, where it rains perhaps once or twice every ten years, and in lush rainforests, where it rains every day. They can be found on Ellesmere Island, where the summer lasts only a few weeks, and on tropical islands, where it is summer all year round. Given the ubiquitousness of bees, how do we obtain accurate comparative information on their diversity in different parts of the world? How do we document bee declines in the same place over the years?

If bees are declining in abundance and diversity, this is best established by looking at them in the same place at intervals separated by a decade or more. How do we sample bees to evaluate changes in numbers over time? The standard method most melittologists use is to walk around and catch them with an insect net. Beady-eyed bee biologists meander across a field site very slowly, looking at the flowers and swiping their nets every time they see something that might be a bee. We don't look only at the flowers, however; it is often easier to catch bees as they fly around their nesting sites. (This is particularly true of the cuckoo bees; because they get other bees to collect the pollen and nectar for their offspring, they spend most of their time searching for suitable nests to plunder.) Having discovered a good spot, a bee biologist will become dreadful company on a countryside walk. On many occasions I have spent hours or even days patrolling the same hundred metres of rich bee habitat looking for undescribed species that I know no one has ever seen before. But this approach is not very scientific; it gets lots of bees, but if people later want to replicate your methods (and it is this repeatability that is the basis of the scientific method), they will find it rather hard to do so. This is partly because some people are simply better at collecting bees than others. They may have better eyesight or better hand-eye coordination; they may be better at telling bees apart from other insects; or they may simply have had more practice at swinging a net. Even the same person will vary in aptitude for bee collecting on different days. If you had a bad night's sleep or are forty years older than when you first surveyed a site, you probably won't do as well.

A variety of methods have been developed to try to make the results of bee sampling more consistent. The most widely used passive bee-collecting device is the pan trap, a small plastic bowl partly filled with soapy water. The soap reduces the surface tension of the water, making it more likely that a bee will sink and drown rather than swim to the edge, crawl up the side and escape. Small plastic pans or bowls come in a wide range of colours, and several people have experimented with different types to see which are most effective. The most common approach is to use a mixture of white, fluorescent yellow and fluorescent blue pans, and to arrange ten of each colour in a cross or a straight line that passes through several different habitats and a wide variety of flower patches. The pans can be put out approximately once a week throughout the flight season, and most of the bee species in that area will be sampled.

The armchair entomologist can also use trap nests to sample bees. As their name suggests, these are artificial nest sites. They can be put out during the winter or early in spring, then collected in the fall once all bee nesting has ceased. The bee brood is then reared. Trap nests are usually blocks of wood with holes in them, like the artificial nesting blocks used for the alfalfa leafcutter bees, or they may be bundles of bamboo canes or some other easily hollowed pithy stem. It is best to make holes of different sizes to entice a greater diversity of bees to nest. In one interesting variety of trap nest, a sheet of clear plastic is placed over a board into which deep grooves have been cut. This design lets us watch what is going on inside the nest without disturbing the inhabitants.

With so many sampling methods available, which should you choose? The answer depends on what you're interested in discovering and how much time and effort you can expend. Putting out pans is a good way of learning what bees are in an area that you do not have the time to survey in person. But which methods are best for finding all the bee species in an area as part of a detailed inventory?

This has actually been assessed, over a whole continent, by the biologist Catrin Westphal of the University of Bayreuth, in Germany. Catrin and colleagues from across Europe sampled a variety of habitats in France, Germany, Poland, Sweden and the United Kingdom. By sampling such a wide variety of places, with varying habitats and climatic conditions, they could be moderately confident that whatever method worked best for them would likely work almost anywhere with a temperate climate. They tried a variety of netting approaches, as well as pan traps and trap nests, and counted all the species found by each of the methods. They then calculated what proportion of the total each individual method had discovered. None of the methods collected all the species at any one site, but pan traps collected the greatest proportion of the total. Not surprisingly, the trap nests, which surveyed only those bees that use holes in wood or stems as nests, did fairly poorly. But they did reveal patterns that were similar to those obtained by all other methods combined. In summary, the data suggested that the best single method for surveying bees is the pan trap, but that even trap nests could be used to compare diversity among sites.

To compare bee diversity on a global scale, we need methods that will work in all habitats. What about deserts and rainforests, habitats that the European team did not survey? Many researchers have found that pan traps work very well in arid environments but not in jungles. However, most of the traps put out in jungles have been set at ground level. Perhaps they would work better in the tree canopy, which is where most of the flowers are.

Simon Potts, the lead pollination biologist at the Centre for Agri-Environmental Research at the University of Reading in the United Kingdom, has studied how best to sample bees in tropical rainforests. He and his colleagues have tried all sorts of trapping methods in the jungles of Africa, and they have solved the problem of getting traps high up into the trees. Using catapults and crossbows, it is possible to loop a rope around the highest tree branches, tie small platforms with pan traps (or other sorts of traps) to the rope and then string them up at almost any height. Simon has faced incredulity from his university's financial administrators, who wonder why his research grants are being used to purchase weapons of small-scale destruction. But the results should be sufficient to sway even the most skeptical of deskbound university bureaucrats on the worthiness of such unusual expense claims.

~

Several of the patterns we see in bee biodiversity we know with greater certainty from studying other living things. The most impressive of the patterns we have discovered is the latitudinal gradient in species richness—that is, the gradual increase in the number of species as you move towards the equator from either the North Pole or the South Pole. This pattern has been attributed to everything from average temperature, variance in temperature and the surface area of the land to whether the area was once glaciated. The problem is that many of these explanations are interrelated, and teasing apart the various potential causes is complicated.

Despite the controversies, some explanations seem unequivocal. For most organisms over much of the surface of the earth, the amount of available energy is a major determinant of the number of species an area can maintain. There are numerous ways of measuring available energy, but for our purposes, we'll consider it to be a combination of warmth and sunshine. Warm, sunny weather works very well as an explanation of patterns in both plant and animal diversity for most of northern North America (in practice, this means Canada plus a strip of land just south of the Canada-U.S. border). It is the best explanation of variation in species richness of organisms as divergent as trees, butterflies and birds. This means that if you were to count the number of tree or butterfly or bird species in the squares of an imaginary grid placed across the surface of Canada and Alaska, you would find that the species richness increased as the amounts of heat, sunshine or any other energy-related variable increased. In the comparatively cooler regions of North America, the number of species found in an area is limited by available energy, which on average increases as you move south.

We can understand why by using an economics analogy. It is easier for people to specialize in comparatively obscure jobs in a rich economy than in a poor one. In a well-off part of the world, individuals can earn their living as abstract painters, fruit fly geneticists, brain surgeons or melittologists. In a poorer economy, it is more difficult to earn a living in any of these endeavours. Similarly, greater specialization may be possible in a region of the globe where there is lots of energy for creatures to use than it is in regions with less. The result is that a greater number of species can coexist in parts of the world where there is more warmth and sunshine.

The problem with this hypothesis is that it breaks down completely in warmer parts of the world. In North America, energy seems to be a limiting factor only for the number of species north of a line joining Boston to Seattle. So what determines variation in warmer areas in most of the continental United States?

Throw a grid over a relief map of North America and count the number of different colours found in each of the squares. For a square that falls on Florida, the variation in altitude will be very small and the area of land under the grid will likely all be dark green. A square of the same size thrown over Delaware (the state with the lowest highest point) would also be one colour, plain green. But look under your grid when it is over parts of Colorado and you will find a wider variety of colours, ranging from yellows along the eastern border with Kansas, through various shades of brown, to purple and white (an appropriate colour to use for areas that are so high up and cold that glaciers persist).

In most of the United States, the number of colours under your grid will correlate nicely with the number of species of organisms found in that square. This means that the determinant of large-scale biodiversity for most of the continental U.S. is the variability of habitats, or what is called *habitat heterogeneity*. This is conceptually not difficult to comprehend. In a mountainous area, there will be some species adapted to low-altitude conditions, some adapted to mid-altitudes and some that like it at the tops of the mountains. Some of these plants and animals will be restricted to narrow altitudinal ranges (yes, there are bee species that live only above the tree line in the Rocky Mountains). Species that are adapted to live on mountaintops will not find suitable habitat in Florida or Delaware, but they will find large areas of suitable habitat in Colorado.

Are there areas where habitat heterogeneity ceases to become the limiting factor? Yes. In tropical regions, water availability becomes the limiting factor. The Amazon is not known for its mountains, but it is known for its phenomenal species richness, its heat and its humidity.

These analyses are based upon organisms for which accurate distribution data are available, such as trees, birds, mammals, reptiles, amphibians, butterflies—the kinds of organisms that interest lots of people. The patterns do vary somewhat among these groups, however. For example, the species richness of reptiles is greater in the U.S. southwest than it is in the southeast at the same latitude (and also at the same altitude). For reptiles, heat is more important than water as a limiting factor; they are cold-blooded and need warmth to become active, so reptiles are more diverse in the hot, dry parts of the continent. But for amphibians—which are also cold-blooded but need water to breed in—it is a combination of energy and water that determines the number of species.

But what of bees? No formal analyses for them have been performed yet. This is partly because, inexplicably, bees are not as popular as frogs, butterflies, birds and lizards as study animals. Also, as we've seen, bees are difficult to identify, and identification is a necessary prerequisite if we are to understand how many species are in an area. There are no Peterson field guides for the identification of bees (yet). Much of the information we need to replicate the research already performed on the better-known groups of animals can be found in museum and university collections, but little of it has been entered into the enormous databases that are required for the analyses to take place (although this work has started).

This is disappointing, but there is a shortcut we can take. Ask some bee fanatics where they would like to go for their next field trip and you might be surprised to hear that few of them would say the jungles of Africa, Asia or South America. Especially if offered limitless funds and a guarantee of personal safety, they would most likely choose to go to the Atacama, the Gobi, the margins of the Sahara, Namibia, Ethiopia, Central Asia (the various "stans" that used to be part of the USSR) or the Middle East.

When we consider what we do know about the global distribution of bee species richness some interesting patterns emerge. We know, for instance, that bees are more diverse in warmer and drier areas (with the exception of absolute deserts, where there is insufficient rainfall to permit any vegetation at all). The single most bee-diverse couple of hundred square kilometres on the planet—at least that we know about so far—is in the Sonora Desert on the Mexico-Arizona border. Robert Minckley, an ecologist at the University of Rochester, has found in that desert over five hundred species—that's approximately twice as many as are found in the whole of the United Kingdom, which is ten times larger than Bob's study area. California has over sixteen hundred species, more than twice the number found in Canada, despite the fact that California is less than one-twenty-fifth Canada's size.

The highest levels of bee diversity in North America are found in Arizona, Nevada and southern California, far to the south of the line where energy ceases to be a limiting factor for the animals and plants we have a better understanding of. Why are bees so diverse in deserts? One reason is that the proportion of bee species that specialize on collecting pollen from one or a few closely related plant species is greater in semi-arid and arid regions than it is in damper ones. This empirical finding has been accepted for decades. But why is this true? Again, we have to wave our hands around and make suggestive arguments because we have no experimental verification.

∼

The bee course at the American Museum of Natural History's southwestern research station is held in late August or early September because that's a good compromise between Arizona's late-summer rainy season, which brings new bee-friendly flowers, and the start of the university term in September. One of the students in the 2007 course was the Indian melittologist Hema Somananthan. As we drove from the airport in Tucson to the research station, I spotted a rainstorm in the mountains to the north and mentioned to Hema that we were currently in what the locals called the monsoon season. She laughed. "No, really," I said. "It is called the monsoon season here." Coming from India, Hema thought of monsoon season as torrential rains over much of an entire continent for weeks on end.

In Arizona, by contrast, it is possible to stand on a mountainside during the monsoon season and remain bone dry while several thunderstorms crash in the hills around you.

Deserts are areas of high variation in rainfall. In any one location, it may rain buckets one year and then not at all for several more years. Some places will be awash in colour as the desert literally blooms. But the same place the very next year may be utterly devoid of flowers. Because individual bees can forage to collect food for the next generation for only a short period of time, it is essential for them to initiate nests and start collecting pollen for their offspring when there is enough food. This is more difficult to predict in a desert than it is somewhere with more consistent weather patterns, such as a prairie or a tropical forest. So the bees must use the same environmental cues for their emergence as the flowers use for flowering.

Different plants need different amounts of rain to produce flowers. Going to the same desert over many years to participate in the bee course has demonstrated this to me quite forcefully. In some years there will be a sea of orange-coloured Arizona poppies along the road between Portal and Rodeo, but in others there will be none at all. The amount of orange here seems independent of the amount of yellow on the creosote bushes that flower in the same area. Small plants have different rainfall requirements than large ones, and large plants with shallow roots have different requirements than the deep-rooted creosote bushes.

Research has demonstrated that the growth and flowering of different plant species in the same desert will be triggered by different environmental cues, and that even identical environmental conditions will result in different plants developing somewhat asynchronously. The bees have to adapt to this uncertainty, and one way they can achieve this is to track the requirements of one particular type of flowering plant. If the bees and their flowers can respond to similar environmental variables, the plants will flower when female bees are available to pollinate them, and the bees will be searching for floral resources just as their preferred plant species comes into bloom. If the bees are not cued in to particular host plants, they may miss all potential food sources. But how does this translate into more species of bees?

Imagine a landscape with patches of different flower species in different places. A generalist bee, which can collect pollen from a wide range of species, will be able to obtain sufficient food more or less throughout this region. But a specialist bee, which collects pollen from only one or a few related flower species, will reproduce only within commuting distance of its preferred floral host. Specialist bees are likely to be forced to persist in more fragmented populations than generalists.

With fragmented populations, there is, by definition, little exchange of individuals—in other words, reduced levels of migration from one population to another. The less common migration is among populations, the more genetically differentiated from one another those populations will become. In the terminology of genetics, populations that are strongly connected through lots of migration are said to have high levels of gene flow—individuals flow around the landscape, taking their genes with them. Populations that are fragmented have lower levels of gene flow—individuals don't move their genes around much, and local populations become more differentiated from each other. Specialist bees should have low levels of gene flow because there is less movement of individuals among their populations. Generalist bees should have higher levels.

Bryan Danforth is an enthusiastic bee researcher based at Cornell University and the world's leader in reconstructing evolutionary trees for bees based upon DNA sequences. For his Ph.D., he studied a particularly complex group of bees that form a whole swarm of species mostly in Mexico and the U.S. southwest. There are over eight hundred described species in this group, and most of them are floral specialists (some identification keys refer to the plant the bee was collected from rather than the structure of the bee itself). Bryan studied levels of gene flow among populations in one of these species, the Portal macrotera (a small black-and-red bee with a pale face and a wide head named after the village in Arizona where the bee course is held each year).

To observe the behaviour of his chosen bee species, Bryan spent every day for weeks on end in a box buried in the desert soil. Many researchers have observed similar ground-nesting bees in laboratories, where the bees can be encouraged to nest in a thin sandwich of soil between two glass sheets. But Bryan studied his bees by going underground with them—an unusual level of melittological dedication to duty.

To study the degree of interconnectivity among populations of the Portal macrotera, Bryan collected samples from a range of locations in Arizona and New Mexico. He found that these bees persist in highly isolated local populations with comparatively little gene flow among them. Some of the populations were separated by only a few kilometres, but the level of gene flow among them was as low as it is among populations of some other bee species that are separated by distances of over one thousand kilometres. Bryan's little bees are real homebodies.

This gives us an explanation for the increased species richness of bees in deserts. First, the bees have to be cued in to the unpredictable flowering, which depends upon unpredictable rains. Second, it may be advantageous for bees to coordinate their responses to unpredictable rainfall with those of a particular species of plant. Third, once specialization has evolved, reduced gene flow is an inevitable consequence. Fourth, an inevitable result of reduced gene flow is increased speciation rates. Of course, all this may be exacerbated by geographic patterns in rainfall—bees forage every year here, once every two years over there and so on. This makes it even less likely that one species will exchange genes with bees in other populations that were, originally, of the same species but living at some distance away.

One of the reasons I was stuck in the desert at the beginning of this book was because I wanted to perform a study similar to that completed by Bryan Danforth. I was sampling one specialist bee, the red-bellied leioproctus, and a generalist, the greyish colletes, at exactly the same locations. Although I caught all the bees on the same species of flower (the tricoloured loasa), the red-bellied leioproctus collects pollen only from that plant, whereas the greyish colletes uses a wide range of plants and is common even a thousand kilometres south of the southernmost range of that flower.

The results of this study showed substantially reduced levels of gene flow in the specialist compared to the generalist, presumably because the latter could persist all over the landscape, where any one of many different flowering plants was found, whereas the specialist could survive only in the more restricted areas where the tricoloured loasa was present.

Simply put, species-level divergence is expected more often among specialist bees than among generalist ones. The flow of genes through the landscape actually prevents geographically disparate populations from diverging into different species. With very low levels of gene flow, differences accumulate, and eventually these differences become so great that the isolated populations become different species. So desert bees are more likely to diverge into multiple species over extended periods of time. This is likely a positive feedback loop: species that are already specialists generate even more species of specialists.

∾

Most regions of most countries are heavily influenced by human activities—we've farmed, we've built towns and cities, we've logged forests. So where do we go to find areas of high bee diversity in regions of the globe that have been transformed by human activities? How about a nice sandpit? Not exactly a pristine setting, but remember that bees are more diverse in semi-arid environments. What could come closer to desert conditions than a nice sandpit?

One of the most surreal experiences I have had surveying bees was in a sandpit in Kent, in the southeast of England. The faint buzzing I heard on approaching the pit became louder as I descended the slope to the flat bottom. The noise was deafening by the time I was on flat ground. I started feeling almost seasick. The ground was seething with bees; my eyes wouldn't focus.

The pit was the site of an aggregation of yellow-footed mining bees. Innumerable females were flying around, low to the ground. Those that had started nesting earliest were returning home, their hind legs covered in pollen. Others were leaving their nests and flying in loops around the entrance in an orientation flight, memorizing the surroundings to be able to locate their nests on their return. Those that were not among the first to emerge from their wintry sleep were trying to find a suitable nest site. Vast numbers of males were also flying around, searching for mates.

Doubtless there were other species in the pit as well, but it was difficult to find them among the seething masses of yellow-footed mining bees. I retreated to the edges of the pit, where the mind-numbing hum and visual confusion caused by the sheer number of bees was less disorienting.

The conservation of small sandpits may be essential for the survival of large numbers of bee species in an otherwise agricultural landscape. Although the yellow-footed mining bee is found in other areas where there are sandy soils, it is possible that the species wouldn't endure without the occasional sandpit dotted around the countryside. Ecologists refer to this kind of demography as a source-sink population structure. Vast areas of land inhabited by a particular species may actually not be especially good for it. For example, areas where pesticides are occasionally sprayed during the bees' flight period may serve as population sinks (regions of negative population growth). Bees may arrive there in one year and start a small population, only to be wiped out by an agricultural chemical sprayed at a different time of year than usual. Such sinks cannot persist indefinitely without replenishment from source areas (sources of immigrants that may recolonize sink areas after the populations there have disappeared). In the example of the yellow-footed mining bee, the sandpit would serve as a source and the surrounding agricultural land might serve as a more extensive sink. If we removed the source, then the species might cease to survive over a wide area.

The need for small areas of unusual habitat within an ecologically different matrix is an example of habitat heterogeneity, which as we've seen is often important in determining species diversity at continent-wide scales. But habitat heterogeneity is

The green burrowing bee (*Ctenocolletes smaragdinus*, of the Australian endemic family Stenotritidae) is an astonishingly fast flyer. Not surprisingly, Terry Houston, who has caught more of these bees than anyone else in the world, is also a keen tennis player.—*Claudia Ratti and Nicholai de Silva*

This masked bee (*Hylaeus*, of the family Colletidae), pictured in my garden, is relatively unbee-like. The species, which nests in raspberry canes, does not look much like a bee, mainly because it carries pollen back to its nest in its stomach rather than on hairs on the surface of its body.—*Amro Zayed*

This solitary mining bee (*Andrena*, of the family Andrenidae) is pollinating a raspberry flower in my garden. As the common name suggests, each female makes her nest in the ground, alone.
—*Amro Zayed*

This beautiful bee is a female bicoloured agapostemon (*Agapostemon virescens*, of the family Halictidae). These are communal bees; multiple females live in the same underground nest, but without a queen or worker castes. The males look very similar to the females, but the males' abdomens have black and yellow stripes. These bees are very common in southern Canada and the United States.—*Amro Zayed*

This is a long-legged oil-collecting bee (*Rediviva emdeorum*, of the family Melittidae), which uses its front legs to get oil from the floral spurs of twinspur flowers.—*Kim Steiner*

Unlike other bees, females of the leaf-cutter and resin bee family collect pollen only on long hairs on the undersides of their abdomens. This photo shows the pugnacious leaf-cutter bee (*Megachile pugnata*, of the family Megachilidae).—*Theresa Pitts-Singer*

This is a male long-horned bee (*Melissodes*, of the family Apidae). These bees can often be found asleep on sunflowers on late-summer evenings. The females have normal-length antennae and can be seen collecting pollen from the same sunflowers earlier in the day.—*Amro Zayed*

This female cuckoo bee (*Triepeolus*, of the family Apidae) has no pollen-collecting apparatus. Instead, it lays its eggs in the nests of long-horned bees. Cuckoo bees still need to visit flowers to get nectar for energy to fuel their stealthy operations. —*Amro Zayed*

also important at smaller scales, and its increase over time seems to have been the cause of a rise in bee diversity in one of the most intensively surveyed small areas of the planet.

One of the main gaps in our understanding of bee diversity is a lack of surveys conducted at the same place over a long period of time. One such temporal comparison was undertaken by a student of mine, Jennifer Grixti, who in 2002 and 2003 repeated a detailed bee inventory that had been done in the 1960s. In the original study, three different sites had been surveyed, all in a particularly attractive region northwest of Toronto. Thirty-five years later, Jennifer wanted to survey these same areas again. One was now the site of a monster home, and the exact spot where the earlier survey had taken place was covered in asphalt, bricks and concrete (not an appealing site to survey bees, even if permission could be obtained to do it). A second site was now the home of a clay-pigeon shooting club. Not unsurprisingly, she was not granted permission to walk around on club grounds. The third site was more promising. Although a home had been built on the land, much of the area around it remained quite natural. So with the permission of the owners, Jen repeated the survey at this one site.

We knew what we expected to find. Given the ever-increasing human impacts upon the environment, we expected fewer bee species than in the earlier survey. Given global warming, we expected a reduction in the number of species that were adapted to colder climates (these species should have moved north). We even thought we might catch some species never found in Canada before—species that had moved north from the United States now that climatic conditions were slightly warmer. But we found none of these things.

Using exactly the same methods at exactly the same site for a similar amount of time as in the previous study, Jennifer found that the number of bee species had increased substantially, from 105 to 150, over the thirty-five intervening years. She found no evidence that the species gained were more southerly, warmth-loving bees, or that the species lost were more northerly. We were initially at a loss as to how to explain these data, but eventually we decided that an increase in complexity of the habitat had permitted the persistence of more bee species.

At the time of the initial survey, the site was an abandoned field surrounded by forest, with a few flowering shrubs, lots of flowering herbs and some bare patches of soil. In Jennifer's survey, a lawn had been added, as well as some flower beds, and these had richer soil than the bare patches of the earlier survey (some of which were still in evidence). The number of flowering shrubs had increased, providing lots of food for bees that preferred those particular floral hosts. Perhaps bee species that liked dense grass had been able to take advantage of the lawn maintenance activities of the new owners, while those species that preferred nesting in bare sandy soil still had adequate sites available. Bees that liked nesting in twigs and holes in wood were also better off because now there were more shrubby plants with dead stems that were warmed by the sunshine during the day. The best explanation for our unexpected result was a local increase in the complexity of the habitat.

Although we do not have much in the way of hard data, it certainly seems clear that, all other things being equal, a somewhat disturbed habitat will have more bees than one that is undisturbed. A pristine forest will not have many bees, while a forest that is disturbed by fields and pathways will have more bee species, especially those that nest in bare ground exposed to the sun. In the tropics, rainforests have their own special bees, those that like orchids or flowering trees high up in the canopy. Stingless bees are common in rainforests, often to the apparent exclusion of most other groups of bees. But forests in general do not promote very high bee diversity. Low-intensity agricultural land, with patches of woodland, roadside verges and occasional bare soil along a footpath or in a sandpit—these are areas likely to harbour a greater diversity of bees.

The areas where I have found the greatest diversity and density of bees are the rural regions of the Dordogne in France and some of the hillsides in the disputed territory currently in northern Israel. The Dordogne is dominated by low-input agriculture, with an intermediate level of disturbance and little pesticide use, and it provides a patchwork of different habitats that shift over time. I could survey large numbers of bees in the Dordogne just by sitting down next to a patch of dandelions. I could count more bees in a few minutes than I could find in most other habitats in an hour.

Semi-arid land, such as that found in the Golan Heights and other parts of the Middle East, has similar characteristics to somewhat disturbed habitat in damper areas as in the Dordogne: patches of bare soil provide the perfect nesting substrate for burrowing bees, and many dried stems house twig-nesting bees.

Perhaps surprisingly, urban areas often provide a complex habitat that can be very good for bees. I would guess that there are at least one hundred species to be found in downtown Toronto; there are more than thirty species in my own small backyard. But there would be fewer species in the newly constructed suburbs. It is not difficult to imagine why. My backyard has a wooden grape arbour where large carpenter bees nest, raspberry canes for small carpenter bees and masked bees, and soft brickwork for orchard bees to burrow into. While a ground-nesting bee that was active only late in summer would find it impossible to negotiate the dense vegetables in my backyard, there are plenty of gardens nearby that would be just fine for them (because most of my neighbours are better gardeners than I am and do not crowd their plants so much). This explains why my squash plants are pollinated by hoary squash bees that tumble into and out of the large yellow flowers early in the morning. I have never seen them nesting in my garden; they must commute to my zucchini from someone else's backyard.

Some of the bees that nest in the walls of my house, the wood of the grape arbour, the bamboo poles that I use to string up the tomatoes or the old raspberry canes left over from the previous year's crop might not find their preferred floral hosts in my vegetable-laden garden, but they can find them somewhere else in the neighbourhood.

If you want to find an ecologically complex single square kilometre, you would be hard-pressed to improve upon an old city's downtown neighbourhoods, assuming that the inhabitants are more interested in gardening than they are in covering their property in concrete. Every backyard shows the signs of different planting preferences, and the overall results may be spectacularly diverse.

~

By now, you doubtless consider me a fanatic who wants as many bees as possible to be found in as many places as possible. This is generally true, but there are exceptions. The wool carder bees that engage in territorial battles in my garden shouldn't be there. They shouldn't be in New Zealand or Brazil either. This species is native to Europe, but it found its way to North America and many other parts of the world presumably because of its penchant for nesting in holes in wooden structures such as packing crates.

Numerous other bees are on continents where they do not belong. These out-of-place bees have been taken outside their native range by us, either on purpose or by accident. Species that are present in areas outside their native range are often called invasive. But for an alien species to be truly invasive, it has to be shown to have a negative influence on the living things in the region to which it has been introduced. We rarely know this for bees because of a lack of surveys before and after the arrival of the introduced species.

We have been taking other species around the world with us ever since we started moving from place to place. Red foxes were taken to Australia by aristocratic Europeans who wanted to continue the age-old barbarism of fox hunting in the new continent. The local marsupials were not impressed; some died out, and some persist only on fox-free offshore islands. Rabbits are a well-known pest in Australia. The same folks who brought those foxes put a lot of effort into introducing rabbits to the large, previously rabbit-free, antipodean landmass. They had to try really hard to get the rabbits to take to the continent, and succeeded only on the third attempt. Rabbits now compete for grassy resources with two agriculturally important introduced species, sheep and cattle (or lesser and greater grass lice, as some Australian conservation biologists call them). Rats were inadvertently taken throughout the globe on ships, and they often had a devastating impact on the birds and mammals native to the countries they were introduced to, including Australia. Domesticated cats that are allowed to roam outdoors cause similar ecological devastation in areas where aggressive predators were not previously found.

Many invasive species are real pests. Worldwide, they cause billions of dollars of damage each year. The problem with invasive species is that they are self-propagating: once they have arrived, they generally increase in numbers and spread as widely as their climatic tolerances permit. Pollution is a breeze to fix in comparison. Turn off the source of the pollution and the problem will go away. The source of invasive species, however, is travel and trade. We are not likely to stop travelling, and we are not going to stop trading. But even if we did, the invasive species that have already been introduced will keep on reproducing and likely keep on increasing their geographic ranges on the invaded continents. Once they're established, turning off the source does nothing to impede their spread.

Are there any examples of bees being moved to a foreign location and having a negative impact on organisms in their newly occupied territories? The answer is yes. The honey bee, for example, which is native to Africa and parts of Europe, has had a devastating impact in Australia. People have found native wild Australian birds and marsupials stung to death because they were trying to live in places where they once nested undisturbed—all because the alien honey bees decided that a hollow gum tree was the perfect new home for them.

There is some evidence that a large leafcutting bee—the giant resin bee, introduced into North America from East Asia—is taking over from the Virginia carpenter bees, because it usurps the nests of the homegrown species. It is probable that many other introduced bees are competing with native bees for floral resources, but we cannot say for certain because we generally do not have good data on what was going on before the introduction.

It is not a good idea to move bees—or any other species, for that matter—into areas where they were not previously found. Such shipment of bees has resulted in considerable economic damage to us. Parasites and diseases of honey bees and bumble bees have been moved from continent to continent and wrought havoc. If the varroa mite were transported to Australia, the almond growers of California would probably be incapable of obtaining enough hives for their crop in an emergency. If another new disease emerges in honey bees or bumble bees, or is transmitted from a wild species of bee to a managed one, and then spreads worldwide, much of our food supply could be put in serious jeopardy.

Bees can be found in almost every environment, from deserts and jungles to cities and sandpits. This makes bees valuable indicators of the quality of almost any type of terrestrial habitat. But whenever an area is good for bees, it will also be good for the natural enemies of bees. The ways in which these enemies go about their dastardly deeds is fascinating and the topic of the next chapter.

10

Anti-Bees

SEXUALLY TRANSMITTED CHILD-EATING FEMALE IMPERSONATORS

ON A CALIFORNIA SAND DUNE

In the entomology course that I teach every other summer, I do a couple of sessions on beetles. Since it's true that most organisms are insects and it's almost true that most insects are beetles, I figure they deserve at least two lectures. These come towards the end of term, by which time the students have realized that, occasionally, I am a little eccentric. When I tell them that they are about to see what I consider to be the most exciting piece of video footage ever taken, they are far from sure what to expect.

I show the students a video taken by Leslie Saul-Gershenz, a conservation biologist with a special interest in insect biodiversity and chemical ecology. To make this video, Leslie had to spend innumerable hours sitting on a sand dune watching insects. What the video shows is a stick with a brown blob at the end. From stage left, a fast-flying object zooms towards the blob, making contact with it, then exiting stage right. The blob is gone. After a pause, the camera pans to the right to show a large bee writhing on the ground. Eventually the poor beast becomes airborne again, rapidly exiting the frame at the top. Mind-numbing or extremely exciting? You can judge for yourself by watching the video at the journal website (see Sources at the end of the book). But I guess that your answer to my question will depend upon how fascinated you are prepared to be by (somewhat anthropomorphized) rat-sized, sexually transmitted parasites that impersonate the females of their hosts to become giant child-eaters. I guess that needs some explanation.

What this video footage shows is one step in the lifecycle of the Francisco oil beetle, a large blue-black flightless beetle that looks kind of greasy (hence its name). It is related to the Spanish fly (another kind of oil beetle, not a fly at all), which has been used by some as an aphrodisiac. The Francisco oil beetle is a natural enemy of the pale-footed habropoda, a large, robust grey-brown bee. What Leslie filmed was the transfer of recently hatched beetle larvae onto a male bee. To become an adult oil beetle, the larva has to end up inside a nest where a female pale-footed habropoda is storing pollen and nectar for her own offspring. Upon gaining entry to the nest, the beetle larva eats the pollen masses—and sometimes the bee's offspring also. But how do the larvae get from a male bee into the nest? And why do the larvae transfer onto a male bee, when the male isn't collecting pollen and nectar, and he isn't going to take the juvenile beetles home to a food-filled nest? Finally, why does a male bee make contact with a brown blob of beetle larvae in the first place, especially when it is likely to result in the death of his offspring?

Male bees are randy little buggers whose sole aim is to mate with females that will use their sperm. They try to mate as often as possible (in this way, they are not very different from males

of most other species). But with their fast flight speed and eyesight that is better suited for detecting movement than seeing
details of shapes, this can be a recipe for disaster. If you go into
your garden when male bees are flying—which will be almost all
spring, summer and fall—you can see them pouncing on almost
anything that vaguely resembles a female of their own species.
You might see a male bee land on a different bee species (male or
female), a wasp, a fly or even a dark necrotic spot on a plant leaf.
Clearly these insects have their little brains so addled with sex
that they don't always pay attention to details. And that's good
news for the beetle larvae, which have evolved a remarkable strategy to attract these male bees to them. They produce a blend of
odours very similar to the chemical mix that the female bees emit
to attract mates, and in so doing, they trick the males into "thinking" they are the objects of their desire.

With our comparatively poor sense of smell, it is not easy for us
to understand that male insects can often locate females from considerable distances simply by odour alone. Insect antennae are elab-
orately ornamented with innumerable little structures that permit
them to detect tiny quantities of specific chemicals. The chemical
blends that insects use to communicate with one another are called
pheromones. Some male insects can detect a single molecule of a sex
pheromone, and by zigzagging as they fly upwind, they can trace
those pheromones to their fecund source. They can travel many kilometres to home in on a single pheromone-emanating female.

In Leslie's video clip, the male pale-footed habropoda saw
something that was the same size as and had a similar smell to a
female bee, and he acted accordingly. He approached it, made contact or attempted to mate, and all of a sudden, he was covered in
beetle larvae. This explains why the male bee approached the baby

beetles in the first place and also why it was a male, rather than a female, bee that made the initial contact.

These nest parasites are sexually transmitted. If that sounds rather icky, it gets worse when you imagine that the interaction between bee and beetle larvae is akin to you chatting up some attractive person at a bar and then finding that he or she has dis-integrated into seven hundred rat-sized parasites that attach them-selves to your back. You are then forced to take the parasites home, where they will proceed to eat your children.

But how do the beetle larvae get from the male into a female's nest? They will transfer to a female when the larvae-covered male mates. Given the mating propensities and visual acuity of male bees, the larvae often get the opportunity to transfer to other males when a male mistakes a him for a her. The beetle larvae actually increase the chances of such inadvertently homosexual behaviour because while hitching a ride on a male they continue to produce the same chemical blend as the female's sex pheromones. Now not only do they smell like a female bee, but they are sitting on a male bee and making him smell like a female while he flies around acting like a bee. The beetle larvae are transferred to other males as well as to females. Sometimes the beetle larvae are often forced to act like long distance hitch-hikers, picking up many rides before reaching their final destination.

The female beetle lays a clutch of seven hundred or so eggs beneath plants that grow on dunes. After the larvae hatch, they migrate upwards and wait at the end of a stalk or an experimentally placed stick, where they give visual and chemical cues to suggest to male pale-footed habropoda bees that, rather than being a bunch of beetle grubs, they are actually a female bee. When a male bee comes near enough, the mass lunges towards him, and if contact is made, the entire mass of hundreds of larvae transfer onto him. How they transfer, en masse in a split second, I do not understand. But once they've transferred, the male bee is so smothered he's almost unrecognizable. The weight of the larvae causes the bee to instantly crash to the ground (but as we saw in the video, he eventually becomes airborne again).

This is an extreme example of the lengths to which some natural enemies will go to access the large amounts of food gathered in a bee's nest. Even most other species of oil beetles are not as devious as the Francisco one. Some lay eggs directly on the flowers that bees visit; their larvae only have to hop onto a female bee as she forages on the flower. The larvae of other oil beetle species also do not cooperate with each other—the trick played on the bees by the Francisco oil beetle larvae would not work if they were each acting alone. This is an interesting and unique evolutionary phenomenon.

Female bees are such good providers—they make a home, collect all the food required for their offspring's complete development and then close the door and leave. This makes for quite an easy life for a successful interloper, so it is not surprising that many organisms have evolved strategies for gaining access to this horde of food.

Some bee flies employ an interesting tactic: the females hover over mining bee nests and flick eggs down the entrances. But before doing this, they sit in the dust for a while and suck soil particles into their rear end. The soil covers the egg, making it largely indistinguishable from a loose piece of dirt on the wall of the bees' burrow. In this way, the female bee fly camouflages her egg, which she then aims down the nest entrance like an expert sharpshooter.

The bee fly story is not over yet, though. The dust-covered egg hatches inside the nest and becomes a small, active larva; this then moves around inside the burrow until it finds an open brood cell that contains a pollen ball under construction. It waits on the wall of the brood cell until the mother bee has done her job, completing the food mass and laying an egg. It then burrows into the pollen ball and waits some more. After the bee larva has hatched and almost finished eating the pollen, the fly larva latches on to the poor defenceless grub with its mouthparts and sucks it dry. This strategy gives the bee fly a juicy larva to eat rather than the comparatively dry and grainy pollen. And by waiting until the last minute, it ensures itself of the largest meal possible (unlike tender lambs and tough mutton, bee larvae do not get less tasty as they grow, but like sheep, they do provide a larger meal when older). Still, the bee fly larva sometimes has to burrow through the ground to find a second grub to suck dry to complete its growth.

Many of the things that cause mortality in bees have nothing to do with chemical mimicry or ovipositional sharpshooting. Bees that nest in stems may have their entire brood consumed by birds that split open their homes and eat the row of developing brood inside. For ground-nesting bees, a thunderstorm may be more of a problem. A heavy rain can waterlog the soil, causing the pollen balls to go mouldy, or soak the larva, exposing it to a microbial disease. (There are reasons why your mother told you not to go outside in the rain when you were young.) Bees have a variety of strategies to minimize such dangers. Many species waterproof the inside of the brood cell with glandular secretions, which they paint on as a lining. At extremely high magnification, you can see the brushstrokes made by their tongues. It used to be thought that these linings maintained humidity inside the brood cell, but preventing the entry of water is likely just as important. Some bees waterproof the pollen ball rather than the brood cell. Others make little three-legged stools out of the pollen ball so as little as possible is in contact with the soil beneath.

The pupae of at least one bee species have the capacity to swim—something quite unusual for the stage in the lifecycle that is generally characterized by immobility.

Bee larvae are probably not well equipped to deal with diseases, but they have evolved strategies to help them escape some potential illnesses. One involves the co-evolution of bees with tiny mites that crawl over the larvae eating potential disease-causing microbes. The mites act as hygienic helpers, just as an egret cleans a water buffalo's hide of ticks and our parents removed nits from our hair. Females of some bee species have special rosettes of hairs at the base of the abdomen to carry the mites, which arrange themselves over these hairs like shingles on a roof. The most extreme mite-storing structures are found in some African species of large carpenter bees. Some house mites on the anterior surface of their first abdominal segment; there the mites are protected by the adjacent surfaces of the body, which develop to surround the mite-bearing surface. But some species have gone even further, creating a cavity in the abdomen with only a tiny, mite-size opening to the outside. The mites are completely surrounded by the body of the bee. Once transported to the host nest, they emerge and run around cleaning it of vermin. This would be something like our having a stomach-sized cavity in our chest or abdomen in which we carry kittens when we move house so mice can be removed from our new dwelling. I love my two house cats, but I would never want to carry them around internally.

Not all mites are beneficial, of course; we've already seen that tracheal and varroa mites cause considerable economic difficulties for honey bees and their keepers. Wild bees also suffer from blood-sucking mites that feast on their bodily juices.

❧

Perhaps surprisingly, some of the worst enemies of bees are also bees—the cuckoo bees. The world's expert on the biology of cuckoo bees is Jerry Rozen, curator emeritus at the American Museum of Natural History in New York and organizer-in-chief of the annual international bee course. In addition to enjoying a gin and tonic or two during happy hour, Jerry has a particular predilection for discovering the nests of bees so rare that few melittologists have ever seen them. But the even rarer cuckoo bees that lay their eggs in the nests of these bees are Jerry's main quarry. He wants to find out what crafty methods the cuckoo bees use to gain entrance to the host nest, and to learn how they prevent the hard-working host bee from finding their eggs and destroying them. Consequently, Jerry spends a lot of his time digging holes. This would be an energetic task for a young bee biologist, but Jerry is over eighty and can still dig a bee burrow faster than anyone. The fact that he often does this towards the end of a long day pacing back and forth along a patch of ground looking for the nests is even more impressive—and may explain the need for those gin and tonics at the end of the day.

Like their hosts, cuckoo bees come in a variety of sizes. Some are a mere two millimetres long, and their eggs are obviously much smaller than this. Using painstaking excavation techniques, Jerry manages to find these tiny eggs—although they're almost indistinguishable from the sand grains on the wall of the host bee's brood cell. Some larger cuckoo bees nail their eggs into the wall of the host's brood cell so only the extreme end is exposed. Others secrete a film over their eggs so they don't really look like eggs at all. Cuckoo bee eggs are placed in precise locations that vary among species; some even lay their eggs directly on top of the host egg, which then becomes the cuckoo bee's first meal upon completion of its embryonic development (cuckoo bee eggs often develop faster than their host eggs just to ensure that they can do this).

Cuckoo bees employ a wide variety of techniques for gaining entry into host nests. Some use chemical mimicry, so the host bee does not detect them as an enemy. Some are stealthy, creeping in while the host bee is away foraging or waiting until the nest is finished and then gaining entry after the host bee has left for good. Some are built like tanks and use brute force to fight off the host's attempts at repelling them. A cuckoo bee may even enter host burrows with its sting raised over its head like a sword, ready to attack anything that stands in its way with both mandibles and venom.

The best example of a detailed war strategy I know of among the bees involves a cuckoo bee called the long-horned sphecodes and its social host, the messenger sweat bee. The interactions between these two species have been studied in most detail by the Austrian-Canadian entomologist, and my own Ph.D. supervisor, Gerd Knerer.

The messenger sweat bee lives in colonies of up to a hundred workers. The amount of nutritious food kept below ground in just one nest is enormous, so it is not surprising that many enemies want to obtain some of this booty. But the nests of this species are well defended because there are often dozens of workers capable of biting and stinging any intruder. The nest's structure also facilitates defence. Like the castles of old, the entrance and passageways are so narrow that only one individual can pass at any given time. The nest entrance is also just wide enough that a guard bee can completely block it with its head. Indeed, the entrance is so small that the queen can't get out even if she wants to (queens are substantially larger than workers in this species). As if this isn't enough, the nest entrance is lined with a ring of cement-like material made from a mixture of excavated dirt and glandular secretions produced by the bees.

The way the cuckoo bee gains entry to these fortresses reads like a lesson in ancient military strategy. The females wait until dawn on a warm morning following a night of rain. Rain softens the hard-baked mid-summer soil, making it easier for them to dig into. Rather than attacking the nest directly at the entrance, which is defended by the cement-like reinforcements and a permanently positioned guard, the cuckoo starts digging down towards the burrow from a short distance away. This serves two purposes: it gets the cuckoo into the burrow beneath both the sentry at the entrance and the hardened lining, and it isolates the guard, which can be killed from behind if necessary. The cuckoo still has to deal with the nest's remaining workers, which come to the colony's defence. The cuckoo female will encounter these one at a time in the narrow confines of the nest burrow, and she kills or maims each one. The dead and dying bodies impede her progress towards her objective—the pollen balls lying deep inside the nest—so she drags them to the soil surface and throws them out. Arriving a few hours after dawn, the melittologist will find a scene reminiscent of medieval battlefields: little piles of vanquished corpses and the bodies of dying attackers litter the ground. The cuckoo bees usually win the battles because they are well built for attack and well defended against their host. They have scythe-like mandibles and wide heads to give extra strength and purchase to their bites. They have heavily sculptured bodies, with pits, ridges, flanges and grooves that deflect or catch the stings and mandibles of the defending hosts. Like knights of old, the cuckoo bees have soft regions separating the segments of their armour, without which they would be incapable of movement. The pits and ridges on the surface of their bodies ensure that the host's weaponry is directed away from these weak links in their physical defences.

Fighting its way into a nest with hundreds of inhabitants is difficult enough, but it would seem an impossible task for a bee to invade a colony containing tens of thousands of individuals. This is precisely what the robber bee does, however. Robber bees are stingless bees that take all their nesting materials and food from the nests of other species of stingless bees (and occasionally from honey bees). The species of this bee that has been studied in the most detail is the lime robber bee. Once a suitable victim nest has been chosen, a few of these shiny black bees attack. These are often killed by the host bees, but the attackers are able to release copious amounts of a lemony-smelling alarm pheromone while the initial foray continues. This brings more attackers to the scene. Eventually, dozens to hundreds of thieves invade the host nest through a mixture of brute force and chemical warfare. The lemony odour is so strong it overpowers the chemical communication system of the attacked bees, who become so alarmed that they hide in the inner recesses of their nests, leaving their construction materials and food supplies for the attackers to plunder. Once the invasion is well under way, the robber bees defend the attacked nest to prevent the return of foraging host bees. They will also modify the nest entrance, and will sometimes even take over the entire nest and move in.

Given the havoc wrought by robber bees, it is not surprising that some of their potential victims have developed defensive techniques. For instance, some have evolved an early warning system that helps prevent them from being targeted. Individual robber bees "scout" out nests suitable for attack. The potential victims often station a dozen or more worker bees outside the nest entrance to watch for robber bee scouts and other enemies. If they detect a scout—evidence that their nest is being considered for attack—they repel it or kill it. This ensures that the information that their nest may be ripe for plunder does not get passed back to the marauders waiting in the robber bee nest. Other species will pour honey over the invading robbers, just as the defenders of medieval castles did with hot oil. The honey gums up their enemies, rather than scalding them, making it hard for them to move.

~

So far, we have concentrated on enemies that attack the nests of bees. But there are plenty of hazards awaiting adult bees as they forage. Many bees fall prey to flower spiders, sit-and-wait predators that are coloured to match the flower they sit on. Some of these spiders are like chameleons; they can change their colour at will. If you see a bee in an unusual position at the edge of a flower, it has likely fallen victim to one of these predators and is dangling from its killer's jaws. I've been humbled to find spiders in possession of bee species I haven't found after an entire day of wandering a field. This is perhaps a measure of how much better spiders are at catching bees than the average melittologist is.

One of the more interesting attackers of adults bees are wasplike flies called conopids. These flies lurk around flowers looking for victims. They make brief contact by jumping onto the back of a bee and then darting away. During that moment of contact, they insert the barbed end of an egg into the soft tissue between two segments of a bee's abdomen. The barbs anchor the egg to the bee in the same way that a hook lodges inside the mouth of a fish. The fly larva hatches from this egg, burrows around inside the bee and starts eating the contents of its abdomen. The larvae of some species of conopids have a long neck, and for a long time nobody knew why. It was Gerd Knerer who accidentally discovered its purpose while studying the enemies of the messenger sweat bee.

Having run out of insect preservative in which to pickle bees, Gerd decided to kill some by putting them in boiling water instead. Death was instantaneous for both the bees and the conopid larvae inside them, and as a result, the long neck of the parasite remained in place, permitting Gerd to learn what it was for. We already knew that these dastardly grubs feed upon a bee's abdominal contents. The abdomen contains reproductive organs, fat stores and the digestive system, and the bee can survive for a while without these. The story would be quite different if the conopid fed on the head or the thorax first; without flight muscles or a brain, the bee would not last long at all. (If you incapacitate your living food, you increase the chances that it will be eaten by something else, such as a shrew or a spider. And if the animal you are eating from the inside is swallowed whole from the outside, you won't survive the event any longer than your host will.) So Gerd accidentally discovered that the conopid's long neck permits it to gain access to the flight muscles through the narrow waist between thorax and abdomen. This gives the conopid one last, highly nutritious meal before the bee dies and the fly larva pupates inside the abdomen of its dying host.

We have already met another important natural enemy of bees, the bee wolf. There are actually over thirty species of these wasps, and different species hunt different types of bees in different parts of the world. One of them, the bumble bee wolf, catches only bumble bees as food for its offspring. Aggregations of these wasps can number in the thousands of nests, and each has a female wasp intent on paralyzing bees. She has to immobilize perhaps half a dozen of them to provide enough food for the complete growth of a single offspring.

Reuven Dukas is a member of the department of psychology at McMaster University in Hamilton, Ontario. He has studied the learning process in organisms as diverse as blue jays, fruit flies and honey bees. From 2002 to 2004 he studied an aggregation of bumble bee wolves in Wyoming and the ecological impact of these wasps on the area around their nest site. He calculated that the wasps sometimes killed as many as a thousand bees each hour. Their impact was so great that the wasps had to continually increase the area over which they hunted. At the beginning of the season, the wasps hunted on flowers immediately around their nests. But by the end of their foraging period, after a few weeks of depleting the neighbourhood of bumble bees, the wasps had to concentrate their efforts on patches of flowers that were almost four kilometres from their nests—that's a long way for a wasp to carry a bumble bee. Reuven estimated that over the course of about one month, the wasps at the nesting site he studied depleted bumble bee populations over an area of approximately fifty square kilometres.

But the impact of the wasps extended beyond the bees. Reuven suspected that by depleting their pollinators, the wasps might have an indirect effect upon the reproductive success of the local plants, and he was right. To prove his point, he counted the number of bumble bees visiting patches of flowers at different distances from the wasp aggregation. Bumble bees visited coneflowers outside the hunting range of the wasps fourteen times more often than those within it, and goldenrod twenty-six times more often. Monkshood proved particularly interesting. Before the wasps became active, the number of bees visiting patches of this plant close to where the wasps nested was very similar to the number visiting a patch five kilometres away (outside the maximum hunting range of the pred-atory wasps). But towards the end of the wasp's annual hunt, the number of bees visiting the flowers within the hunting radius had dropped by 50 percent in comparison with those outside the range of the wasps. This reduced the reproductive success of the plant. Outside the area affected by the bumble bee wolves, approximately 40 percent of monkshood flowers produced seeds; inside, less than 20 percent produced seeds. This is suggestive of a cascading impact through the ecosystem. It also demonstrates the importance of pol-linator abundance for flower reproduction.

The honey bee wolf we met on the Isle of Wight at the begin-ning of the third chapter can build up dense nesting aggregations that are typically home to several thousand wasps. This size of predator population can result in the deaths of thirty thousand honey bees (half a colony) in a single day. Not surprisingly, hives kept near aggregations of honey bee wolves are not always eco-nomically viable for the beekeepers. But the impact of the bumble bee wolf is greater than that of the honey bee wolf. Wiping out one thousand bumble bees in an hour is equivalent to losing the entire workforce of maybe a dozen of the largest colonies.

Still, the impact of bee wolves is nothing in comparison with the impact of the number-one enemy of many of the world's bees. We see this organism every time we look in the mirror. Humans are driving species to extinction at a rate unseen since a huge meteor hit the planet just off the coast of the Yucatan Peninsula sixty-five million years ago. That event triggered the fifth mass extinction on the planet. We are having an equally meteoric impact upon the planet, and we're causing the sixth mass extinction of all time. In the next chapter, we'll see exactly what we're doing to the bees.

11

What Are We Doing to the Bees?

BEE-FREE DAY IN GERMANY

I n August 1986, I travelled to Munich, Germany, to attend an international conference of researchers who study social insects. There were whole symposia about ants, termites, wasps and bees. I was there to talk about some of the barely social bees that have just a few workers in each nest. I arrived a day early, and since the sun was shining and it was nice and warm, I decided to wander the countryside looking for bees.

I took a train to the edge of town and walked among the fields looking for bees on the weeds along the roadside. By eleven o'clock I was perplexed: I had seen nothing other than a couple of honey bees. Perhaps it had been cold the previous night and the bees were feeling drowsy, or maybe it had rained in the night and they were cleaning out the debris that had washed into their underground burrows. But by one o'clock, the only wild bee I'd seen was one ruddy driveway bee, and I was no longer making excuses for the bees' lack of attendance to duty. Clearly this was not a good area for bees at all.

If this had been a part of rural Kent in England or the Dordogne region in France, or even the agricultural belt around Toronto,

I would have seen at least a dozen species of bees by that point. I knew something must have made that part of the German countryside particularly bad for bees, and I guessed that it was pesticide use. This suspicion gained support when I developed a runny nose later that evening. I surmised that pesticides might be the cause of both the absence of bees and my bothersome nose.

One of the earliest studies of the impact of pesticides on bees was carried out by Peter Kevan, the pollination biologist we met in Chapter 9. Peter became heavily embroiled in the pesticide issue after being contacted by New Brunswick blueberry growers who had noticed an alarming reduction in the yield of their crop. New Brunswick is mostly covered in trees, and the health of the forests has long been threatened by the spruce budworm, a pest moth that defoliates and kills trees over large areas. The predominant method to control the insect was aerial spraying with pesticides. Blueberry growers in areas adjacent to sprayed woodlands soon noticed a large decrease in the productivity of their crop.

Peter set out to investigate whether the spraying was responsible for the crop losses. During the flowering period, he vigorously waved a sweep net over the blueberry bushes. On fields distant from the spraying he caught many bees, but he had less luck in pesticide-affected areas. On average, he found four times as many pollinators far away from the pesticide area; intermediate locations had three times as many pollinators as the areas close to where the pesticides were sprayed. The reduced abundance of pollinators persisted for several years after a spraying.

Peter also sent the bees he collected away for chemical testing. The bees from fields close to sprayed forest had, on average, more than twenty-five times as much pesticide in their bodies as the bees from far away. This strongly suggested that spraying for spruce budworm was resulting in a large-scale reduction in bee populations, and that this in turn was reducing the yield in the nearby blueberry fields and causing a loss of income for blueberry farmers. The farmers launched a lawsuit against the forestry industry and won. The spraying was discontinued—at least close to the blueberry farms—and the bee populations, the blueberry yields and the farmers' financial well-being all recovered.

This is one of the best-documented examples of human impact upon bee populations. It is also a story with a happy ending. Nowadays, pesticide-free buffer zones are commonly established around blueberry fields as a result of the litigation.

I don't think humans have always had a negative effect on wild bees. Indeed, it is quite likely that for most of our history, we have had a mutually beneficial relationship. By clearing forests for agriculture and making pathways through fields and woodlands, we have created habitats that have often been better for bees. Footpaths and roadside and railroad embankments provide an open, sunny and warm substrate for ground-nesting bees. Before we came along, those ground-nesting bees were restricted to places like riverbanks, areas cleared by fires, the dirt exposed by a tree being uprooted by the wind and naturally occurring bare soil patches. Our activities likely resulted in increases in bee populations, and the bees, in turn, pollinated our developing crops. At least in north temperate regions, human activity traditionally had a positive impact on bee populations. But the pendulum has now swung in the other direction. Thanks to us, many bee species are on the verge of extinction.

The imminent extinction of pollinator species has been catalogued by the Xerces Society, an organization aimed at promoting the conservation of insects. (It's named after the Xerces Blue butterfly, the first North American insect species known to have become extinct.) In 2005, the Xerces Society published a list of the endangered pollinators of North America. Fifty-one species of bees were listed, and eight of them were thought to be already extinct. But these were just the ones a handful of experts could think of at short notice. Since then, there have been surveys done in the Las Vegas area, where some rarer species could not be found despite considerable effort by some of the best bee observers I know. It seems likely that two species may have become extinct, and both are known only from single holotype specimens. These losses are certainly just the tip of the iceberg; many species that we thought were safe when we were first asked to compile the list are now, less than a decade later, much rarer.

There are many bee species that have been seen only a handful of times. Often these are small insects living in remote parts of the world that have gone unnoticed by entomologists even if the regions have been visited by them. These undiscovered species likely number in the several thousands. Quite a few have undoubtedly been wiped from the face of the planet over the past two hundred years or so without our having noticed that they were here in the first place. But there are also bees that, like Lazarus, seem to have returned from the dead. (Of course, they were here all along—we just hadn't seen them for a while.) One example of this is the macropis cuckoo bee, a small dark brown bee that lives in damp places. This species had not been seen in fifty years when Cory Sheffield (the world's expert on the bees of Canada and a research associate in my laboratory) rediscovered it in Nova Scotia, a part of the world where it had not previously been recorded. Alas, repeated searches in the same place have failed to detect it again. This species is likely to be one of the first added to any nation's endangered species list. To gain a better understanding of the disappearance of bee species, though, we need to seek out those for which we have more reliable data—those that are more easily observed and are (or at least used to be) less rare.

Bumble bees are among the most noticeable and charismatic of all invertebrates. They are large and brightly coloured. Their gaudy appearance is a warning to would-be predators that they are well armed with a powerful sting. Because they advertise their ability to retaliate, bumble bees do not expect to be attacked and so are more readily approachable than most bees. You can often observe them up close without disturbing them. They are also active all spring, summer and fall. This combination of bright colouration, ease of observation and long activity period makes bumble bees prime subjects for scientific investigation. Consequently, serious analyses have been made of their populations in many parts of the world, and declines have been documented in North America, Europe and China.

In North America, it was the disappearance of Franklin's bumble bee that first aroused the attention of bombologists (as we sometimes informally call bumble bee experts), and one in particular. Robbin Thorp is a robust, jovial, retired bee expert based in California. He looks a lot younger than his years (doubtless because of the happiness induced by studying bees in California for so long), and he is also one mean pool player. He first found Franklin's bumble bee in the 1960s. In 1998 he initiated serious surveys for the species, and very soon he noticed that it was declining precipitously. Franklin's bumble bee has traditionally been found in southern Oregon and northern California, one of the smallest geographic ranges known for any bumble bee. Because of its distinctive colour pattern—yellow fur on the head and the front two-thirds of the thorax and the rest of the body black—it is easy to identify this species in the field. But for three years in a row, Robbin didn't see a single one. In the next year, 2006, he saw one individual, but his subsequent searches drew blanks until 2009, when he found it once more. If Franklin's bumble bee isn't extinct already, it is certainly about to become so.

But Franklin's bumble bee is not the only species in trouble in North America. The appropriately named western bumble bee used to be abundant from southern California all the way up to Alaska. In recent years, people have had great difficulty finding it, although some remnant individuals have been seen at the northern and southern extremes of the bee's previous range, and it can still be found at high altitude in Colorado. In eastern North America, the rusty-patched bumble bee has disappeared from most of its historical range, Ashton's cuckoo bumble bee seems to have disappeared completely and the yellow-banded bumble bee (an uninspired name, since most bumble bees have one or more yellow bands) also can no longer be found in many places where it used to be common.

The declines of these species were abrupt, suggesting that something fairly dramatic must have happened. Perhaps not surprisingly, that something seems to be associated with humans. What is surprising is that the human activity that was the likely cause of these declines is related to our need for bumble bees.

Bumble bees are commonly used to pollinate greenhouse crops. They are social bees, but unlike honey bees, they start a new colony each spring with a single queen. It takes a while for these colonies to develop a sufficient workforce to be able to pollinate a greenhouse crop, and producing these large colonies in a factory-like setting has become a small industry. Wild caught or domestically reared queens are kept in large numbers, and the colonies they produce are then shipped out to greenhouses that require them. A mature colony costs approximately $175 (not including shipping and taxes). Unfortunately, it was discovered, after the fact, that the bumble bee colonies being supplied by these industrial-scale rearing barns were almost all infected with one or more bumble bee diseases—but perhaps not North American bumble bee diseases.

Early in the development of the North American branch of this bee-rearing business, two species were taken to Europe for trials. One was native to eastern North America, the jewelweed bumble bee, and one was native to the west, the aforementioned western bumble bee. It is suspected that when these New World bees were being studied, they picked up some Old World diseases from European bumble bees that were being kept in the same rearing facilities. Perhaps a European strain of a particular protozoan parasite, a kind of bee-infecting malaria, was introduced to North America when the bees were returned to their native range. These diseased bees were then shipped out to greenhouse growers. If the bees had been kept inside the greenhouses, things might have been okay, but the indoor bees often got outdoors to forage on flowers on the other side of the glass. One study showed that almost three-quarters of the pollen collected by bees that had their colonies inside a greenhouse was obtained from flowers growing on the outside. This gave plenty of opportunity for the greenhouse bees to transmit their illnesses to native bees outside.

James Thomson is a biologist who has had a long interest in bumble bees. Research performed in his laboratory at the University of Toronto has shown that bee diseases are far more prevalent closer to a greenhouse where industrially reared bees are being used than they are farther away. The results of this study were published under the whimsical title "Plight of the Bumble Bee," but the results were anything but funny. It is possible that the decline of the rusty-patched, yellow-banded and western bumble bees, Ashton's cuckoo bumble bee and maybe even Franklin's bumble bee was caused, or at least exacerbated, by the spread of diseases from domesticated colonies.

But once again, things are not as simple as this. Why is it that not all bumble bee species have been affected by these diseases? Surely all bumble bees should have declined in those parts of North America where greenhouse growers use farmed colonies. But many species remain unaffected. Perhaps this can be explained by natural variations that make some species less susceptible to infections; after all, not all people are susceptible to the same illnesses, and canine distemper will make your pet dog ill but leave your domesticated cats unaffected. In the same way, different bumble bees might be expected to respond differently to the same illness. The fact that the species that are declining in North America are all very close on the bumble bee family tree suggests that they might be influenced in a way that more distant relatives are not. But perhaps something else may be implicated in these population collapses.

The decline of some eastern North American bumble bees is being investigated in great detail by my Ph.D. student, Sheila Colla. Sheila came to my laboratory specifically to find out why the rusty-patched bumble bee was disappearing. First, she repeated a thorough survey of bumble bee diversity that had taken place near Toronto in the 1970s. In the earlier work, the rusty-patched bumble bee made up 14 percent of all individuals seen. Sheila spent vast amounts of time looking for bees in the same areas surveyed in the original study, but she failed to find a single individual. Three years of repeated sampling yielded not one rusty-patched bumble bee, although over one thousand individuals of other species were seen.

Next, Sheila revisited as many as possible of the places where this species had been seen in the past. She visited all sites in Ontario where more than five rusty-patched bumble bees had ever been collected, and she also visited the parts of the eastern United States where it had been seen most commonly in the 1900s. Four years of searching, from 2005 to 2008, for this once common species yielded a stunning total of one individual. This lonesome male was found at Ontario's Pinery Provincial Park in 2005. In 2009, students in my laboratory surveyed the park intensively and found three individuals, all in the same small area. The park now sells T-shirts with an image of this bee on it and the caption "The last stand." It may indeed be the last stand, at least in Canada.

Sheila was not the only person to notice that the rusty-patched bumble bee had become difficult to find. Several people in the United States had also started looking for it. As of late 2009, the species seemed to have disappeared entirely from its previous strongholds in Ontario and the Appalachian region of the northeastern United States. It has been found occasionally in the past few years in Pennsylvania, Illinois, Indiana and Wisconsin, but it has disappeared from vast tracts of North America where it used to be common. Will this species come back from the brink of extinction? Only time will tell.

The decline of the rusty-patched bumble bee took place in the late 1990s. Could the spread of diseases from greenhouse stock have been the cause? This was the initial suspicion, but the timelines did not match very closely. So what other environmental changes might have impacted these bees?

Interestingly, a new class of pesticides was introduced to North America just prior to the disappearance of the bee. The very same pesticides had been banned in parts of Europe precisely because of their effect on honey bees, and beekeepers in several countries were involved in lawsuits against the pesticide companies. The chemicals had been tested on a North American bumble bee species for toxicity and were found to be safe when used according to the guidelines. But there are reasons to suspect that this might not have been a sufficiently rigorous test.

First, different species may respond differently to the same chemicals. The species tested was quite distantly related to the rusty-patched, western and yellow-banded bumble bees, and it's possible that these species were just less susceptible to the same chemicals. Findings based on testing done on a common species might not apply to the bumble bees that are now disappearing.

A second problem relates to the way in which pesticide testing generally takes place. Pesticides are designed to kill things. Some concentration of a chemical is sprayed on the target organism's body or added to its food. A research technician then simply records what proportion of individuals do not survive and compares that to a control group that was not exposed to the chemical but was sprayed with a placebo. This is fine and dandy for checking the lethality of the substance on the target pest, but it's not so good at revealing non-lethal effects that can have just as negative an impact upon a population.

For example, these pesticides often influence the foraging behaviours of bees. Suboptimal foraging means the bees bring less food home and the colonies don't get as large, and this in turn means that fewer of the males and queens needed for the next year's colonies are produced. But other non-lethal effects may be far worse. What if the bees survive when they ingest or otherwise make contact with these chemicals, but their reproductive organs are destroyed? If you sterilize all individuals in a population, you will wipe out the population as surely as if you'd killed them.

What if further testing determines that new pesticides were indeed the cause of the declines of some bumble bees? The ramifications of this would be enormous. If it's proven that the same chemical can have profound effects on some pollinators but no apparent effects on others (even if they are closely related to those impacted), it will mean that the dangers of pesticide use are more complex than previously imagined. So how extensive should the testing of a pesticide company (or preferably an arms-length independent organization) be? If a chemical has a catastrophic effect on only four of forty bumble bee species, does this suggest that each new chemical should be tested on at least ten species? And that's just bumble bees. What of other species of bees or other beneficial organisms, or other organisms that may or may not be beneficial?

~

Pesticide use is not the only thing reducing the number of species on the planet. The three anthropogenic impacts that conservation biologists think are the most important are climate change, invasive species and habitat fragmentation.

Some predict that because of climate change alone, up to 60 percent of the planet's species will, over the next fifty years, be unable to find enough suitable habitat for their populations to be viable. As the climate warms, cold-adapted species have to move north (in the northern hemisphere, or south in the southern one) or go uphill to remain within their required climatic envelope. A mountaintop species in Arizona isn't likely to be able to move farther uphill as the climate warms, and nor is it likely to survive a move across many kilometres of lowland desert in an attempt to find a higher mountain. Of course, the more mobile a species is, the more likely it will be able to move to areas that have its preferred conditions. The estimates of high extinction rates were for the least mobile species. More mobile species are expected to have lower rates of extinction.

As the climate changes, some bees will find they have less area where conditions are suitable for them, but others will have more. As warmth-loving insects, many bees should see an increase in their habitat as the temperatures climb. But habitat fragmentation means that these areas will be farther away and more difficult to reach. In some cases, only the most mobile of species—migratory birds, for example—will be able to find suitable habitat. Bees are not migratory creatures. The smallest species seem the least capable of moving very far. Small bee species may be most at risk of extinction from climate change.

Every time the climate scientists recalculate the rate of warming, they come to more dramatic conclusions. The predictions of up to 60 percent species loss seemed radical when they were first suggested, but now they sound rather conservative. Things are worse than we realized, and there are concerns that we might be causing a runaway greenhouse effect, with climate change setting a vicious cycle in motion. The thawing of peat bogs and regions of permafrost, for example, releases more carbon into the atmosphere and warms up the climate even more.

It remains unclear to what extent the changing environmental patterns will result in a decoupling of the activity periods of bees and the flowering periods of the plants they pollinate. Some studies have predicted substantial extinction rates even in the absence of any other environmental changes, simply because the bees and their flowers will not track each other's needs as closely as they do now. Given what we know about the variation among plants in the cues they use to flower, it seems unlikely that most bees will be able to respond quickly enough to changes in those patterns. I hear an ever-increasing number of anecdotal accounts of fields full of flowers with no bees in attendance. In intensively managed parts of the world, this might be the result of any number of human impacts. But in semi-desert regions and rainforests where human impacts are minimal, climate change may be disrupting the normally close ties between bees (and other pollinators) and the flowering plants that depend on them for sexual reproduction.

We are also responsible for the catastrophes that result from the introduction of invasive species. These are self-perpetuating causes of extinction: once a population has become established, it can increase in both number and geographic range. Aliens have contributed to the extinction of at least 170 animal species, and in 20 percent of those cases, invasives were the only cause of extinction.

Over thirty alien bee species have been introduced into North America. The golden-tailed lithurgus, a western Mediterranean bee that specializes on the pollen of spotted knapweed, is one such species. It was first detected damaging a New Jersey home by burrowing into the wood. The home was close to some railway tracks by the Delaware River, and the bee had probably arrived from Europe in wooden packing crates. For a time, it was thought that the bee had disappeared from North America, but it was rediscovered in 2007, having apparently spread far afield, travelling mostly along the railways.

An invasive species is one that has a negative ecological impact in areas where it is introduced. The golden-tailed lithurgus certainly had a negative influence on that New Jersey home, but it remains to be seen whether it will have a broader ecological impact. The giant resin bee is more likely to be a true invasive species because it competes for nest sites with the native Virginia carpenter bee. In general, we have little information on true invasive bees (as opposed to merely exotic ones). This is partly because of a lack of baseline data on what bee species were present before the arrival of the foreign species. Native bees may be more negatively influenced by invasive weeds than they are by exotic bees.

This became obvious to me when, in 2001, I returned to the site in Ontario where I had performed some of my Ph.D. fieldwork. I had spent most of each spring and summer of 1983 and 1984 sitting on the side of Highway 10, just northwest of Toronto. This was, and still is, an extremely busy highway with lots of truck traffic. I had chosen this site not because of its pleasant scenery and relaxing ambience, but because it had a large nesting aggregation of the ligated gregarious bee. The highway embankment had large areas of bare soil with just a few hawkweed plants here and there, and the ground beneath the basal rosette of leaves of this plant was a favoured nest site for this bee. My first graduate student, Miriam Richards, performed some of her Ph.D. research work at the same site ten years later. But now we cannot do fieldwork anywhere along this embankment because it is entirely covered in crown vetch, an invasive weed that is found along roadsides throughout eastern North America. It's a safe bet that there are many other examples of invasive weeds crowding out native flowers at once popular bee-nesting sites, although as I have said, we have no suitable baseline data on most bee species.

The third major impact we are having on biodiversity is through habitat fragmentation, where once extensive areas of forest, grassland or other wild habitat are replaced by human-modified environments, such as farms. Some entire ecosystem types have almost vanished from the surface of the planet. For example, oak savannah, a kind of tall grass prairie that grew on drier sandy soils in eastern North America, now covers less than 0.02 percent of the area it occupied before the advent of European-style agriculture. It is not only the decline in area that dooms to extinction the organisms that live in these increasingly rare habitats; it's also the distance between the fragments of habitat and the complete unsuitability of intervening land. Small organisms are often unable to make it from one small patch to another distant one simply because of the distances involved. And while large organisms might be physically capable of moving from one small forest fragment to another, they are exposed to considerable risk if they attempt it. Roads, for example, are not very safe places for most vertebrates, and most human-impacted areas are covered in a dense network of them.

~

Our species is impacting the others on the planet in many negative ways. But how many species are likely to become extinct in the near future as a result of our activities? That's a very difficult question to answer for several reasons. First, as we discovered in Chapter 5, different biologists have different ideas about what exactly the term "species" means, and some definitions will as much as double the number of species recognized. Second, even when we decide on a definition, we still have a poor idea of the number of species that coexists with us on earth. Estimates vary by over an order of magnitude! While most biodiversity researchers suggest there are close to ten million species on the planet, others say the number might be over one hundred million.

Estimating how many species will become extinct is clearly impossible if we have no idea how many species live on the planet in the first place. However, at least when it comes to habitat loss, we can estimate the proportion of species that will go extinct. Several studies have demonstrated that as a very rough approximation, a reduction in habitat of 90 percent will translate into a 50 percent reduction in the number of species. This simple formula is often used to make dire predictions about the future of tropical rainforest species (because that's where a large proportion of the planet's species lives, and we can estimate the rate of loss of habitat easily from satellite images). It is predicted that half of the species on earth will disappear as a result of rainforest destruction by mid-century. Perhaps half of the rainforest bees will be gone by 2050.

What we do know is that habitat loss, climate change and invasive species are increasing the extinction rate substantially. The millions of animal species that exist today have evolved over six hundred million years, and we are wiping them out almost as fast as a giant meteor would. There have been five mass extinctions over the course of these past six hundred million years, and they resulted in the extinction of 60 to 96 percent of the species that were around at the time. The human population is causing the sixth mass extinction by emitting greenhouse gases, cutting down forests and otherwise altering habitat, and transporting species to places where they don't belong. Does it matter? Why should we care? Even if the world had only 10 percent of its current species left, it would look the same to most people. Most of us do not see or care about the other 90 percent of species (after all, most of them are insects). Indeed, one of the most terrifying things to conservation biologists is that as our species becomes increasingly urban, fewer people care about the rest of nature because fewer ever experience it. But we should care, even if only for purely selfish reasons.

~

Paul Ehrlich is a retired academic who, incidentally, was one of Charles Michener's first graduate students. Under Mich's supervision he began a career in butterfly biology, and he quickly became an advocate for these most charismatic and well known of all insects. Ehrlich has asked us to imagine a thought experiment roughly along the following lines: Picture being on an airplane with a window seat looking over the wing. At thirty thousand feet, you notice that one of the many, many rivets that hold the wing together has broken. This is a bit nerve-racking, but there are numerous rivets left, so you are not overly concerned. Then another rivet pops. You bring this to the attention of the flight attendant, and she tells you that while she doesn't think it matters, she will ask the pilot anyway. Another rivet pops off. The pilot comes back to tell you that the stress-bearing parts of the wing have almost twice as many rivets as they actually need to function. Reassured, you fall asleep. Several hours later, you look out the window again and notice that even more rivets popped during your slumber. As you chew your way through a raspberry-flavoured piece of plastic sponge, a few more rivets go, and the wing falls off. End of you. End of story. End of thought experiment.

The point of this "popping rivet" analogy is that in most complex systems—airplanes, national economies or ecological communities—there is built-in redundancy. An airplane wing, for example,
can lose some of the parts and everything will continue to function
normally. But there is a threshold. If the loss exceeds this threshold,
the whole system may collapse. Airplane engineers probably have a
good idea of how many rivets are needed to hold an aircraft wing
together, but recent developments indicate that economists have
a less precise understanding of what makes national and global
financial systems work. Ecologists generally have even less of an
idea of what the thresholds are for the stability of our ecological
systems. The loss of the rusty-patched bumble bee was certainly
catastrophic for the species that relied on it (including the Ashton's
cuckoo bumble bee that invaded its nests), but we do not know
what impact it will have on other parts of the ecosystems of eastern
North America. Perhaps some plants are now declining as a result
of the loss of this bee. Perhaps some butterflies whose larvae need
those plants as food are now declining. Perhaps those butterflies
were responsible for the pollination of another flowering plant.
Perhaps the fruit from that plant was essential for the survival of
one or more species of songbird. We could reach a situation where
the number of pollinating species falls below a threshold and the
ecological system collapses, becoming less diverse, less aesthetically
pleasing and less ecologically (not to mention nutritionally) productive.

Ecological systems seem to have the capability to flip from one
set of stable conditions to a different set of conditions that may
also be fairly stable. Despite a fifteen-year moratorium on cod fishing in parts of the east coast of Canada, codfish remain rare—they
are not recovering. It seems that the pressure placed on the fish

resulted in large-scale ecological changes that preclude the species from bouncing back. We have altered the oceans to such an extent that the cod now persist at such low levels that fishing for them is a bad idea. Perhaps a loss of pollinator species will cause whole pollination webs to flip irreversibly to less productive systems. In a wilderness area, a loss of pollinators might ultimately result in a decline of the flora and fauna that attracts people to such sites. If the simplified ecosystem is an agricultural one, the loss of pollinators might result in reduced production of tasty berries and other pollinator-dependent crops.

There is good evidence that community-wide reductions in pollinator assemblages have already occurred. People in Great Britain and the Netherlands—where eccentricity among insect lovers (like me) is almost actively encouraged—have been sampling insects for a long time. The data from Britain indicate that a substantial reduction in the number of bees and hover flies (among the most important of non-bee pollinators) has occurred over large areas. Similar declines were also detected for the flora that depends on pollination. In the Netherlands, the bees had declined but the hover flies had not, and so bee-pollinated flowers declined in that country while flowers pollinated by hover flies did not. If similar data were available for other parts of the world, I am sure the same general patterns would be found. It would seem that we are indeed simplifying pollinator diversities, and this may have a dramatic impact on the diversity of natural ecosystems. But what would be the impact on us?

The economic advantages we obtain from nature have been termed "ecosystem services." (I find this term somewhat offensive. It seems to suggest that the whole point of our planet's biodiversity is just to provide "services" for us.) Nature is a bargain. At no monetary cost to us, it recycles our organic waste, cleans the water and helps purify the air, provides fish for fishing and game for hunting, and supplies pollinators for our crops. The economic value of the services nature provides has been calculated, conservatively, at US$33 trillion per year. We don't normally worry about the number of rivets that hold an airplane together, and few of us worry about counting the diversity of nature either. We expect both the airplane and the natural world to continue to function for at least as long as we need them. We notice the importance of forests only when we suffer from flooding caused by the clearing of trees. We notice the importance of ground-covering plants only when the soil erodes and someone's house falls off a cliff. We notice the importance of mangrove swamps only when a tsunami causes devastation farther inland than ever before. Whatever the causes, the wonderful biological diversity that used to exist in many parts of the world is becoming grossly oversimplified, and the consequences will probably not be to our long-term benefit.

Certainly a substantial reduction in pollination would have an immediate impact on our well-being. Some crops are entirely dependent upon pollination, and as I noted earlier, yields of many other fruits, seeds and coffee beans are higher where pollinator diversity is greater. So it follows that a reduction in the diversity of pollinators will reduce the quantity of food available to us. This would not matter so much if the number of people on our planet was substantially less than it is. It would not matter so much if the production of pollination-dependent crops could continue to increase substantially. The agricultural productivity of some countries will not be terribly damaged by a loss of pollinators because these countries primarily produce wind-pollinated grains. But the agricultural sector of many other countries— those places where bee-pollinated fruits and vegetables form a large part of the overall picture—would collapse. These are often the countries where people are already having difficulty obtaining enough food, even though they may actually toil in fields where crops are grown (albeit for export rather than the nourishment of the people living around and working in those same fields).

An additional recent development is further straining our capacity to feed the world's population: our insatiable demand for fuel. The transformation of land from agriculture to biofuel production is resulting in ecological devastation on an enormous scale, and this will probably only increase as oil and gas become more expensive. We need oil and gas to transport seeds to the farms where the crops are grown, and to transport the crops to the people they're meant to feed. We need oil and gas to manufacture, transport and apply fertilizers and pesticides; to pump water for irrigation; to transport the poorly paid farm workers who pick the crops; to process the food. Between the processing factory and your supermarket, the average bite of food will have travelled more than one thousand kilometres. Most people in the developed world eat food that has already used non-renewable energy on over a dozen occasions before it reaches the table. Compare this to the tomatoes that spring up in your garden from several generations of rogue seeds. You use no non-renewable energy walking out your back door to pick these for a salad. This may be why they taste so good.

The problems that agricultural shortages will produce might not matter to the very richest among us, but most people will not be able to afford the foods we now take for granted. The results of this increased social inequity could be problematic. Our perception of our overall well-being depends on where we are in the financial ranks of the society in which we live. The greater the disparity between us and those who are much better off than we are, the less healthy we perceive ourselves to be. What will happen if most people cannot afford fresh fruits and vegetables? Perhaps the unrest that results from increased social inequities will be the real impact of a loss of pollinators.

This loss should be something that concerns us because pollinators affect us directly through our food supply. In the past we have been able to rely on an industrial-scale pollinator, the honey bee, for our industrial agriculture, but we may be unwise to continue this narrow focus and should increasingly diversify our options to other species. We can do this only as long as these other species are available, however, and some are disappearing. It is as if we are reducing our own potential for adaptive change to future developments. Every time a species goes extinct, we have lost a piece of the complex jigsaw puzzle that makes our life on earth possible. Every time a species goes extinct, we reduce our own ability to adapt to an uncertain future. This cannot be good for us in the long term.

With the essential role they play in ecological systems, bees are of above-average importance. But I believe they are also important in an entirely different way: their susceptibility to habitat degradation seems more acute than that of other organisms, and therefore a decline in bees should serve as an early warning signal that bad things are going on in the environment. In the next chapter I explain how and why bees should be used to monitor the state of the planet.

12

The Proverbial Canaries in the Coal Mine

UPSETTING ORNITHOLOGISTS IN ROME

"If all birds dropped dead tomorrow, only chicken farmers and academic ornithologists would be inconvenienced. If all bees died out, there would be worldwide food shortages and perhaps one-quarter of the human population would starve."

That's what I said to an ornithologist, in public, at an international biodiversity meeting sponsored by the Food and Agriculture Organization of the United Nations. (I'm very good at making myself popular with people.) The ornithologist had given a presentation promoting the use of birds to assess the state of the environment. I have nothing against birds, but there are drawbacks to using them to assess most terrestrial habitats. As I was soon forced to admit, though, using birds certainly has some advantages, among them the vast number of people both willing and able to go out and identify and count them. But could bees be good indicators of the state of the world? To answer this, we need to know what it is that makes a good environmental indicator.

The Organization for Economic Co-operation and Development says a good environmental indicator must meet three criteria: political relevance, analytical robustness and quantifiability. Are bees politically relevant? Well, I have never seen a politician campaign based on promises to save the bees, but this may actually happen in the near future, especially in agriculturally intensive parts of the world, such as California, where the economy depends on healthy rates of pollination. In several European cities, beekeepers have already been protesting that pesticides are destroying their livelihoods. So perhaps politicians will have to get involved in pollinator conservation in a serious way soon. Of course, political relevance is ultimately determined by voters. But surely coffee, blueberries and almonds, among other crops, are both relevant and important to the populace as a whole. The role bees play in the production of these essential items is well understood. Without them, we would all be worse off—nutritionally, economically and, with no coffee, probably also emotionally. Without bees, the planet might be able to support only three-quarters of the people it supports today. But we have some public education to do to ensure that bees are considered politically relevant.

What of the other two criteria? Analytical robustness and quantifiability are two sides of the same coin. If you cannot measure something accurately, then analyses of data obtained from it will not be robust. We can assess bee diversity and abundance by wandering around with an insect net or by putting out various sorts of traps, and statistical methods for assessing declines in bee populations are well established.

If we are to use animals to indicate the state of the environment, then we need to choose a group that is sufficiently numerous, in terms of both individuals and species, to provide lots of data. If we want indicators that are useful for global comparisons, we need to choose organisms that are found in all the habitats we want to assess. We need to choose those animals that are sensitive to changes in the condition of those habitats. Bees fit these criteria: they are common in all terrestrial habitats that are not covered in ice or bare rock; there are over 19,500 known species; and the oddities of their sex-determining mechanism seem to make them exquisitely sensitive to declining habitat quality. Add to that the fact that bees are economically and ecologically essential for the continued functioning of just about every terrestrial habitat, and their value as indicators would seem to be undeniable.

Alas, there are two enormous pitfalls: bees are difficult to identify, and as a result, there are few people capable of surveying them accurately. Currently it takes years of practice to be able to identify bees to the species level. But this is changing, and with more general awareness of the role bees play in our lives, this taxonomic impediment can be overcome.

To foment the political will, the public has to become involved. This is happening in many parts of the world, with more and more people getting interested in pollinators. Certainly it is possible to go out and simply count the number of insects visiting a plant in your backyard, but more useful monitoring requires you to identify the species as well. This is difficult, but user-friendly identification guides and semi-automated DNA barcoding can help.

Identification guides should include pictorial representations—digital images of real specimens rather than oversimplified line drawings—of the characteristics biologists use to identify bees. You can find examples of these guides on websites such as Discover Life and the *Canadian Journal of Arthropod Identification*. But not all bees can be identified to species using these tools—at least not yet. Work is under way to make online identification guides available for all parts of the world. This will make the job of bee identification far more enjoyable for far more people. And the more enjoyable it is, the greater the number of people who will want to do it. This then makes it more likely that additional resources will be made available to make this work even easier. This promises to become a positive feedback loop.

The promise of DNA barcoding is that it will make species-level identifications both routine and automated. We already have DNA barcodes from at least one individual for 18 percent of the world's bee species and for more than 50 percent of the eight hundred or so Canadian species (after all, DNA barcoding was invented here). Within a few years, all but the rarest bees in North America should be identifiable from a short piece of DNA sequence. Anyone with a few pennies to spare will be able to send in a small fragment of a bee—as small as half an antenna (they can survive without this quite well)—to find out what species it is.

Of course, DNA barcoding is still expensive at the moment—it costs as much as seven dollars for a routine identification (although this is not particularly expensive in comparison with the costs of training someone to perform the same task using traditional morphological approaches). But the costs will go down with demand, and I hope that the demand will be enormous. We can also temper that demand. Bees that are easily identified can be left alone and only those that are problematic need sacrifice a part of an antenna to the cause. After all, no one with a few hours of training in bee identification would need to remove the antenna from a honey bee, a hoary squash bee or a bicoloured agapostemon to identify it to species level in the field.

We can also reduce the demand for DNA barcoding by narrowing down the range of bees we wish to monitor. There are two ways to do this. First, we can assess which bees are most susceptible to specific habitat changes. Second, we can assess population sizes of different groups of bees to find out whether any specific patterns emerge. Both approaches can be used to determine whether some ecological or taxonomic groups of bees are inherently more at risk than others, and are therefore more likely to indicate the state of specific environments with accuracy. This information can also help us tailor our choice of indicators to the specific environmental disturbance we wish to assess.

~

In the spring of 2004, I was in Santa Barbara, California, discussing the looming pollination crisis with other biologists at the National Center for Ecological Analysis and Synthesis (NCEAS). Also present at this meeting were some of the people we have already met in this book—Taylor Ricketts, Claire Kremen, Robert Minckley and Simon Potts—as well as other melittologists and ecologists. NCEAS is an organization that brings together biologists from all over the world to solve problems in ecology. Given the stresses we humans are placing on the rest of the globe's species, there are plenty of problems for the centre to consider. The place is commonly full with temporary visitors, all discussing and analyzing different topics. Of course, it would be far better for the planet if the problems NCEAS helps to solve were no longer problems and the organization could be disbanded. But I suspect that it will be around for as long as civilization lasts. Indeed, it will likely be necessary for NCEAS to be cloned as ecological problems increase.

The group studying pollinator decline was a large one—twenty-four in all—with experts on a range of areas of relevance to the problem. After initial discussions, we were divided into three groups. The group I was assigned to had to consider what aspects of bee biology might make different species good predictors of the effects of habitat disturbances. Clearly organisms with different lifestyles will respond in different ways to the same environmental perturbation. For example, a solitary mining bee that spends eleven months of the year underground will likely survive a brush fire of the vegetation above. But a bee that nests in dried stems will be cooked by the same event. We had to decide which variables to consider and then code them for all the bee species sampled in a specific set of studies. To qualify for inclusion, a study had to have sampled bees associated with a disturbance such as fire, habitat fragmentation or pesticide use. With this large synthesis, we hoped to discover whether bees with certain traits fared better than others depending upon the nature of the disturbance. These variables are more technically termed behavioural and ecological traits. Our group was called the "traits group," and its members were all "traiters."

Most of us will have noticed that over time, some organisms have changed dramatically in numbers. In some cases, we have a reasonably good understanding of what's happened—a ladybug plague typically follows an aphid plague, for example (ladybugs eat aphids). But we have little overall understanding of what determines population changes for bees, mostly because nobody has gathered the available information to perform a synthesis. Such large-scale syntheses are exactly what NCEAS is for, and we had all brought with us information that seemed pertinent to the analyses we planned to conduct. The centre also has a staff of computer and databasing experts who know how to analyze complex information; all we had to do was decide what data to analyze, gather the information we wanted and put it through various statistical tests.

First, though, we had to determine which behavioural and ecological variables might be predictors of the sensitivities of bees to changes in the environment. The first we came up with was the breadth of a bee's diet—specifically, whether it collects pollen for its offspring from a wide variety of plants or from only one (or a few closely related) species. Specialist bees should be more susceptible to environmental changes that reduce the availability of a preferred floral host, whereas generalists should be able to survive changes comparatively well because they can use almost any old weed as a food source for their offspring.

The second variable was choice of nest site. Bees will nest in an almost bewildering array of places—from bare soil to dense grass, pithy stems to rotting forest logs, and keyholes to snail shells. We decided to simplify matters and look at bees that nest above ground versus those that nest below ground. We knew that some disturbances, such as fire, would more likely kill bees in an above-ground nest, whereas others, such as tilling the soil, would likely have a greater impact on the underground bees.

Next we looked at whether the bee needed assistance from another organism in constructing its nest. While most bees make nests using their own labour, others require the presence of another species. For example, a bee that likes to nest in a hole in a piece of wood may need a beetle to make the hole for it. Anything that has an impact on the beetle will also have an impact on the bee. The greater the number of species a bee relies on, the more susceptible it will be to environmental disturbances.

Sociality was another important variable to consider. Most social bees are active throughout the spring, summer and fall. This means they are exposed to environmental changes for a longer period of time than those solitary species that may be active for only a few weeks. Any environmental disturbance that occurs outside the flight period of a solitary bee may have no influence on it, unless its preferred flowers or nesting sites are directly affected.

Finally, the size of bees is important. Large bees can forage over a greater area and will be more successful at finding food if floral resources are patchily distributed. The largest distance I know a bee to travel for food is eighteen kilometres from nest to host plant. This long-distance commute was performed by an Indian large carpenter bee during a drought. Conversely, some small bees, such as the Portal macrotera, will fly only a few metres for food. One of my favourite bees is a small desert species, Rozen's xeromelissa (named after the world's cuckoo bee expert), which nests in the dead, hollowed stems of the same plant that it collects pollen and nectar from. That's like living, working and shopping all in the same high-rise. Small bees do not have the luxury of going on long-distance shopping trips; they may run out of fuel before finding a good spot. Consequently, we predicted that smaller bees would be more influenced by environmental change than large ones.

These five variables were the ones we "traiters" used to evaluate every species of bee in the studies we'd chosen to review. We also had to know what the environmental disturbance had been. The ones we had enough data for were fire (both recent and more ancient), increased fragmentation of habitat and agricultural practices such as tilling the soil and pesticide use.

There were thousands of entries in the dataset, and over six hundred bee species were represented, some of them multiple times. It took years to collate all the required information (though to be honest, none of us could work full time on this project). For some of the species, all the information we needed was readily available. Others had never been studied in the wild, so we didn't even know whether they were solitary or social. For many of these, we could predict what their trait data would be using their position in the evolutionary tree for bees. For example, no species of the group commonly called solitary mining bees is known to nest anywhere other than in the ground, and none is known to have queens and workers. The common name for the more than one thousand species in this group makes sense, and we could code all of them as below-ground nesters without queens and workers, even though only a small proportion of the species had ever been studied for those variables.

We had data from a total of nineteen studies. Given the number of types of disturbances, this provided only a small sample size for each. But the statisticians on the team were good at squeezing the best results from noisy ecological data. And most of the significant results were fairly commonsensical. We determined, for example, that tilling the soil impacts ground-nesting bees more than it does those that nest above the ground; that social bees are more affected by pesticides than solitary bees are; that increased agricultural intensity disadvantages specialist bees more than it does generalists. But some of the patterns we found could not be so easily predicted. Tilling the fields, we learned, has a larger negative impact on specialist bees than it does on generalists, and increased habitat fragmentation affects social bees more than solitary ones. Some of the results we had expected to find (based on earlier, more narrowly focused studies) proved more elusive. For example, we found that small bees are no more affected than larger bees by isolation.

Although our analyses were based on a limited number of studies, the results strongly suggested that habitat modifications will affect some groups of bees more than others. This means that we can adjust the amount of monitoring we need to do based on the disturbance we're investigating. If we want to know the impact of pesticides, for example, we can concentrate our research on social species like bumble bees. If we want to assess the impact of increased agricultural development, then it would be sensible to study specialist or ground-nesting bees—or better still, specialist ground-nesting bees.

There is another trait that deserves serious assessment: cuckoo bees that lay eggs in the nests of only one bee species should be more susceptible to extinction than their hosts. The cuckoos cannot have larger populations than their hosts, and population size, as we will learn next, is a major determinant of extinction risk. Unsurprisingly, there are many cuckoo bees that are extremely rare, but analyses of their potential for understanding the state of ecological systems are only just beginning.

~

An entirely different approach to assessing extinction risk is to obtain an estimate of population size. This is relevant because smaller populations are more at risk of local extinction than larger ones. The problem with this approach is that bee numbers fluctuate considerably from one year to the next, and this can make it difficult to estimate the long-term average population size of reproducing individuals. Here again we can use genetic methods to help, and it was my attempts to obtain specimens suitable for genetic analysis that caused me to be stuck in the sand in the middle of the Atacama Desert at the beginning of this book. Nowadays, I can preserve my bees in alcohol and assess their DNA directly (or more accurately, I can preserve my bees in alcohol and give them to someone with the skills needed to perform DNA analyses). But back then, such DNA-based methods were exorbitantly expensive, so instead I was using the indirect method of assessing variation in the proteins that the DNA codes for. To preserve the proteins I needed to freeze them, which was why I had that canister of liquid nitrogen with me in the middle of the desert.

Population size is a major determinant of genetic variation—that is, levels of genetic variability are higher in larger populations. But the converse is also true: the more genetic variation there is in a population, the larger the population must be (all other things being equal). Because genetic variation is a relatively easy parameter to measure, it has been assessed in hundreds of different species since the first detailed studies were published, in 1966. It was soon established that bees and their waspy relatives had lower levels of genetic variation than most other organisms. This is because of the effects of haplodiploidy on effective population size. With haplodiploidy, healthy males have only one set of chromosomes. This simple fact immediately reduces the effective population size for bees to three-quarters that for human beings, bears or butterflies. If we have ten males and ten females and all of them reproduce equally, then for a human being, a bear or a butterfly, the effective population size is twenty. But for bees, ten males and ten females provide an effective population size of fifteen, because male bees count for only half. This suggests that levels of genetic variation among bee populations should, on average, be 75 percent that of other organisms—again, all other things being equal. Since other things never are equal, it is not surprising that the data suggest that bees do not average 75 percent the level of genetic variation found in butterflies. In fact, bees average somewhat less, probably because of the inevitable production of some dud diploid males.

Another reason for these lower than expected levels is that many of the species assessed were social, and social bees generally have lower levels of genetic variation and smaller effective population sizes than solitary ones. Why?

Walk around a flower-filled field in late summer and hopefully you will find an abundance of bumble bees. If you find a hundred workers in a few minutes, the population will seem large. But it isn't—all one hundred could have come from a single queen and her single male mate. What seems like one hundred individuals actually counts as only one and a half individuals in terms of the effective population size. Social bees cannot maintain numerous colonies in the restricted area of a single small field because they only start to produce next year's queens and their mates late in summer, having already used up large amounts of pollen and nectar to produce the workers. This places a significant limit on the effective population sizes of social bees; most habitats will not have enough pollen and nectar throughout the spring, summer and fall to allow a high density of bumble bee colonies to exist. This suggests that social bees should be particularly good indicators of the health of terrestrial systems.

The last bee trait that suggests some species might be better indicators of the state of the environment than others again is linked to the Atacama Desert scene at the beginning of this book. I was trying to find out if specialist bees had less genetic variation than generalists. I already knew that specialists made up a larger proportion of the bee fauna in desert and semi-desert regions. So clearly deserts were a good place to collect the appropriate samples. But I needed a desert where the bees were well known, and where the distribution of flowering plants had been minimally changed by human activity. Arizona fitted the former description, but because of long-term grazing by cattle, it didn't fit the latter. So I chose the Atacama Desert of Chile. The bees of this region were well known as a result of studies by Luisa Ruz and the late Haroldo Toro of the Universidad Católica de Valparaiso. They told me which species to study and suggested when and where to sample them.

It was important that we sample pairs of bee species in which one was a specialist and one a generalist. It was also important for each pair to be distantly related on the bee evolutionary tree to the other pairs in the study, and therefore evolutionarily independent. Without this independence it would have been difficult to interpret our results. What would it mean, for example, if we'd found that five specialist solitary mining bees had more genetic variation than five generalist bumble bees? We would not have been able to tell whether the differences were due to specialization/generalization, choice of nest site (burrows in the ground or abandoned rodent nests), sociality (mining bees being solitary, bumble bees social) or some other factor we did not know about. By choosing bees that were closely related within a pair but distantly related to other pairs, we were able to avoid these potentially confounding effects.

My Chilean colleagues and I, along with Robin Owen (a real geneticist), applied to the National Geographic Society for funding. Our application was successful, so we travelled to Santiago and rented two half-ton trucks, each of which was equipped with a large drum for extra petrol, a large drum for extra water and a giant vacuum flask filled with liquid nitrogen. We then headed for the southern portions of the Atacama Desert to collect specimens to test our hypothesis.

The work went well. We collected good samples of many species from a wide variety of locations during the day and in the evenings ate marvellous seafood and drank delicious Chilean pisco (an aromatic brandy) before putting up our tents in the middle of nowhere—usually down abandoned mining roads. But with just one more week of sampling to go, disaster struck. The container of liquid nitrogen in which we had combined all our samples malfunctioned when we were quite a distance from the nearest ultracold freezer. We drove to Valparaiso as quickly as possible and managed to borrow some freezer space from colleagues at a university there. The samples were then placed on dry ice for the journey to Canada, where they were transferred again to an ultra-cold freezer. But it was too late. When I opened a couple of the samples, they smelled bad, a sure sign that they could not be used for our genetic research. Three months of work was completely ruined; it was absolutely heartbreaking. But I did what any self-respecting university scientist would do under the circumstances: I sent two of my graduate students to repeat the work. Because of the Atacama's short flowering season, which depends on the winter rainfall, we all had to wait until the following September before replacement samples could be obtained.

Amro Zayed and Jennifer Grixti were the lucky graduate students who got to share a half-ton truck with a liquid nitrogen container for three months in the southern hemisphere spring of 2002. They had a list of the locations that had been sampled the year before, and they knew which bees to look for and which flowers to find them on. One of the most important flowers is the tricoloured loasa. It is very common in this part of Chile and attracts vast numbers of bees. All you have to do is to swipe your net at these flowers and—bingo!—you've got a bee. But there are two main problems with collecting bees from this plant. First, it is sticky and fragments really easily, so its leaves and stems stick to the net upon impact and make a mess. Second, it stings like hell. On an earlier trip, I had asked Luisa Ruz when the stinging would go away, and she'd said, "Never. I still have some spots that occasionally hurt even though I last touched the plant years ago." So although the plant was common and the bees plentiful, my students had to swing their nets gingerly.

Initially the work went well. Amro and Jennifer were in good spirits when they phoned home to say that they had completed the sampling in the southern part of their targeted range and were heading north. But then things became more difficult. All of a sudden, they stopped finding bees. It was the oddest thing. The flowers looked just as good as they had farther south, but there were no bees for many kilometres—perhaps climate change had induced asynchrony between the bees and their flowers. Day after day, Amro and Jennifer traipsed through the arid scrub finding nothing, and their spirits ebbed. They phoned home again to report that they had not collected a single sample in a week. Knowing that rainfall in deserts is highly variable, I advised them to try their luck even farther north. This hunch was correct. They started catching bees again, and their good spirits returned. Even better, on their return journey south, they found bees at some of the locations that had yielded none two weeks earlier and were able to fill in the large gaps in their samples. As Amro later said, "This field trip was an emotional roller coaster. When it was going well, we were really happy. When it went badly, we got really depressed. But then in the end, it all worked out; the results were very clear, and we ended up elated."

The results, generated after months of genetic research in the laboratory, really were very clear: in each and every pair, the specialist bee had less genetic variation than the generalist. Indeed, on average, the specialists had almost an order of magnitude less genetic variation than the generalists. This is strong evidence that specialist bees persist in much smaller, more isolated populations, which makes it more likely that they will be extirpated following an environmental disturbance that impacts their preferred flowers.

In this chapter I have argued that bees should be used to monitor the health of the terrestrial ecosystems upon which we depend, and that they seem particularly good at indicating the state of ecosystems that have already been modified by us. Fortunately, we do not have to sample all bee species; we can tailor our work to fit the ecological disturbance we wish to assess. Whatever the disruption, there are bee species to help us measure its impact. Nonetheless, it is clear overall that we are having a negative impact on bees worldwide. It's essential that we reverse these negative influences. In the next chapter, I outline some of the things we can all do to help keep the bees.

13

Help the Bees

DODGING HIPPOS AT THE AFRICAN POLLINATOR SUMMIT

"Do you have Peter's cellphone number?" Lucie asked, obviously worried.

"No, what's wrong?" I replied.

"There's a hippo right outside his cabin door, and he should be coming over here for dinner right now. If he walks outside without looking . . ."

In Africa, it's not the predatory lions or the crocodiles or the poisonous snakes that cause the largest number of human fatalities; it's the male hippo that thinks its territory's being infringed or the female that suspects its calf is being threatened. Being trodden on by a hippo is an experience you should definitely avoid—the smaller females weight more than a ton, and a large male can weigh as much as two and a half tons. A few stomps from a hippo will turn you into a two-dimensional stain on the grass. We didn't want this to happen to Peter, who is one of the most important pollination biologists in Africa. Fortunately the staff at the Tala Game Reserve in South Africa was alert to this kind of potential problem, and within seconds a Jeep full of rangers was chasing the hippo into the bush. But we'd had an opportunity to see how fast a hippo can trot. I wouldn't recommend relying on your running shoes to escape from one.

This wasn't the first brush we'd had with potentially dangerous animals during our visit. On the first evening, a family of white rhinos had made some of our group wait for a while before they could get to their rooms. The next night, we disturbed a large male Cape buffalo that was eating the lush vegetation in the watered garden. The animal reared and then stared at us from a distance of no more than two metres. It looked as if it would charge. When the ranger clapped his hands and shouted loudly, the enormous beast took two steps back, but it stayed in attack posture. The ranger clapped and shouted again. This time, the animal took two steps forward, and we backed out of there as fast as we could. I consider myself lucky that slugs are the worst enemies of my homegrown fruits and vegetables. If I had buffaloes to contend with in my garden, I would likely give up; my usual pest-control tactics of sprinkling salt or squishing them between finger and thumb would obviously not work.

We had assembled in South Africa because of the worldwide concern for bees. A summit meeting of bee and pollination experts had been organized by the Smithsonian Institution, the Consortium for the Barcode of Life and the Global Biodiversity Inventory Fund. Local arrangements had been made by Connal Eardley, Africa's pre-eminent bee taxonomist. Forty bee experts from Africa, Europe and North America were sequestered away in this spectacular spot, where, with no easily accessible Internet facilities to disturb us from our work, we were forced to concentrate on bee and pollination issues for hours and hours each day. Large tracts of land occupied by ferocious beasts deterred potential escapees.

Global concern about pollinators has reached a fever pitch. Governments all over the world are beginning to realize that we cannot simply assume that our crops and other plants will be pollinated as much as we need. Indeed, in Kenya there are crops that are producing less than half the amount of food they could simply because of a lack of bees. And in parts of China, people have to go around pollinating orchard crops by hand because there are no non-human pollinators left to do the job. In some cases, it is relatively straightforward to correct this. If a crop is pollinated mostly by large carpenter bees, for example, then the decline in pollination can be reversed simply by increasing the amount of woody brush that these bees use as nest sites. This requires education of the farmers, though, because many are used to relying on honey bees, and some think that all other flower visitors are damaging the crop.

Such misconceptions are not limited to people on the African continent. I was recently interviewed for a Canadian television segment and was asked if bees are pests.

"Pests?" I replied, truly dumbfounded.

"Yes. Some people think bees are pests, so are they?"

"Well, if you don't eat *and* you are allergic to being stung, I guess you could consider bees pests," I replied, trying to sound not too incredulous.

Some important decisions were made over our three days of rhino, buffalo and hippo dodging, although unsurprisingly, many of our major conclusions pointed out the need for more research (this was a meeting of research scientists, after all). Still, many of the suggestions were entirely practical. Several had to do with increasing our understanding of stingless bees. In parts of Africa, the honey of stingless bees is often taken by honey hunters in ways that kill the entire colony. If a nest is found inside the trunk of a tree, for example, the hunters will chop the tree down, killing it and everything that was living on it, including the hive. Other people try to maintain a supply of the medicinally and nutritionally important honey by keeping the bees in a semi-domesticated manner. Unfortunately, few people know how to keep these bees.

Of course, bees aren't in trouble only in Africa. Wherever we have the data to show what is happening to them, bees are in decline. What we need to know is the extent of this decline. Some countries are trying to reverse declines in the number of honey bee colonies being kept by beekeepers. Some have earmarked funds for pollinator research. There are even a few companies that have allocated funds to help us understand more about pollinators and pollination. Häagen-Dazs, the German ice cream company, became sufficiently worried about the bees that pollinate the fruits that go into its products that it started a research fund. (Unfortunately, the company primarily funds research on honey bees, which are not always the best pollinators of the fruits used to flavour its ice creams.)

We are only just beginning to understand what is happening to bees and other pollinators, and more research is certainly needed. Nonetheless, each of us can take simple steps to help preserve pollinators throughout the world. Some of these steps are possible no matter where you live; others apply to anyone with a garden, no matter how small. They do not include demands that you have no children, stop heating or cooling your home, wear only sackcloth and never travel anywhere other than by foot (although all of these things would doubtless assist the conservation of biodiversity). Compared to these other suggestions, the ones I am putting forward are small steps that can mostly be carried out in our own backyards.

1. GROW BEE-FRIENDLY PLANTS, PREFERABLY NATIVE SPECIES

Flowering plants have evolved to facilitate pollination. But horticultural varietals have been dissociated from this need, and this has resulted in spectacularly showy blooms that aren't necessarily useable by bees. Take prize roses, for example. These have pollen-bearing anthers, but they are so densely encircled by rings of petals that it takes an unusually strong, pig-headed or desperate bee to get to them. Gladioli, delphiniums and monkshoods are other flowers that are beautiful to behold but difficult for a bee to handle. More open flowers, such as asters, umbellifers and squashes, have readily accessible floral rewards, and a larger number of bee species can use them.

I encourage you to grow native plants whenever possible. Though not as showy as the horticultural mutants, many of them are beautiful in their own right. Blazing star, goldenrod, sunflower, Joe-pye weed, coneflower, bergamot, evening primrose, cinquefoil, milkweed, willows—numerous plants native to North America are both entirely suitable for garden adornment and beneficial to local bees. Unfortunately, many municipalities have laws against untidy garden weeds, so if you are keen on native plants, you may have to ensure that they are arranged nicely so the enforcers of such anti-bee legislation do not mistake your attempts at bee attraction for illegal carelessness.

If you prefer plants that "work for a living" by providing fruits or vegetables, then most of those requiring pollination will be beneficial to a wide range of bees. Squashes will attract the hoary squash bee. Tomatoes will attract a large number of bumble bees, as well as the red-tipped anthophora—a fat, brownish bee with a furry red tip to its abdomen. Strawberries will attract small carpenter bees and masked bees, among others. Fennel, coriander and other umbellifers will attract masked bees and other small species, including some wasps that will paralyze the pests (such as aphids and caterpillars) that feed on your plants and take them back to their nests as food for their offspring. Various beans and peas are good for leafcutter bees and bumble bees. But my favourite garden plants are raspberries and blackberries; not only because their fruits are delicious and their flowers feed bees, but also because their old stems give some of them a place to nest.

It is also important to offer a diversity of flowers that provide blooms throughout the spring, summer and fall. Bees will starve in a field full of tomatoes with no other floral resources because tomatoes provide no nectar, and without nectar, adult bees lack the energy that fuels their flights. Even the bees that pollinate apple blossoms in orchards benefit from lupines planted around the edges of the fields. This is because lupines provide the resources bees need to keep provisioning their nests after the apple blossoms have fallen. With these extra provisions, their populations can increase in size more easily.

2. PROVIDE NEST SITES FOR BEES

I don't consider it essential that you drill holes into the walls of your house to provide nest sites for bees, but I also wouldn't discourage you if you wanted to do this. Still, there are many nest sites you can provide without reducing the resale value of your home.

Every spring, many people in my neighbourhood put cut raspberry and blackberry canes out on the street for composting. In doing this, they are causing the deaths of untold thousands of bees. Small carpenter bees, masked bees, small relatives of the leafcutter bees and orchard bees—all use old raspberry and blackberry canes as nesting sites and spend the winter inside them. Please refrain from throwing these out. You would be helping the bees by leaving these old canes in place, or even better, by cutting them into foot long-pieces, bundling them together and attaching them to a fence, tree bough or overhanging eave. If you put many potential nest sites together in one place, you will be able to watch the bees go about their daily work of building their nests, provisioning their brood and pollinating your fruits and vegetables. I can assure you this will give you a deeper appreciation of these fascinating insects.

If you stake your tomatoes or other garden plants, try to use bamboo canes to do so. These will serve as nest sites for leafcutter bees, orchard bees, the red-tipped anthophora and other bees that need larger stems than most raspberry and bramble canes. Again, if you become a bee-watching enthusiast, you may want to bundle some bamboo sections together and hang them out where you can watch the bees at work. You will attract a wider variety of stem-nesting bees if you bundle canes of different sizes or put out blocks of wood with holes of different sizes drilled into them.

Rather than bringing these canes and other nest sites indoors over the winter, you should leave them outside. The inhabitants will be used to your local winter conditions, but they will not do so well if their natural cycle is disturbed by the unseasonal warmth of a heated basement or the damp of a garden shed.

If you have wooden benches, fences or a grape arbour, let the large carpenter bees burrow into them—but take care to ensure that the structure remains sturdy enough so as not to cause personal injury. (It will take many years before this is a problem, unless the wood is already soft and rotten for other reasons.)

Ground-nesting bees are not always as easy to attract in large numbers as those that nest in stems. The sloping soil of south-facing rockeries is a highly preferred site for many ground-nesting species. You can watch the solitary mining bees and their waspy cuckoos, the nomad bees, fly around such garden features, especially in spring. The nest site requirements of ground-nesting bees are not well known. One small spot may have hundreds of nests per square metre, while other spots nearby will have none. If you have an aggregation of nests in the dirt of your driveway or in your lawn, you should cherish them and consider yourself lucky. It is also possible to encourage bumble bees to nest in your backyard by building suitable nest boxes for them.

Some other gardening practices are also "bee-beneficial." Gardening experts are never happy to hear this, but not mulching is one—you wouldn't want your house covered in compost, and neither do the ground-nesting bees. If you have to do it, please cover as small an area as possible with these bee-repelling materials.

An edge cut around a flower garden or vegetable patch makes suitable nesting substrate for bees that like starting their nests on a vertical or near-vertical surface. But it is important not to cut that edge back while the bees are active.

If you hoe between your rows of plants, do so rarely, shallowly and as late in the day as possible. If you change the appearance of the ground surface and obscure their nest entrances while the bees are outside foraging, they may not be able to find their way home again. If you hoe in the evening, few bees will be outside their nests. Although they will probably have to refurbish the top few centimetres of the burrow the next morning, you will not have caused their brood to be orphaned.

Do not apply layers of wood chips, pebbles or other surface-obscuring materials. Almost no bees will be able to get through to the soil beneath. If at all possible, get rid of that lawn and especially any concrete or paving stones and turn your outdoor space into a bee-friendly flower or vegetable garden.

3. DO NOT USE PESTICIDES

Some city governments are beginning to pass anti-pesticide laws that will improve your health and that of the animals that live outdoors on your property. Unfortunately, the pesticide companies are not happy, and some are suing these more enlightened town councils. Shame on them.

Certainly there are pests we do not want in the garden, but the chemicals that are used against them are not good for beneficial insects, and some of them are not good for us either. Insecticides often kill off more of the parasitic wasps that are nature's biological pest-removal system than they do of the aphids and caterpillars that the wasps might have controlled if they had not been poisoned. There are far more environmentally friendly ways of getting rid of pests than adding poisons to the environment. An infusion of garlic sprayed on infested plants can often dissuade insect pests from feeding and breeding. Squashing aphids, caterpillars or slugs between finger and thumb is a more labour-intensive approach that may not appeal to the faint-hearted. You can always just pick the pests off and drop them into soapy water or some other drowning agent, such as a little beer or diluted vinegar. Or you could pay a local teenager to do the job for you.

4. BUY ORGANIC FOOD WHENEVER POSSIBLE

We know that some pesticides are lethal to bees, and it seems probable that others render them sterile. We also have evidence that organic farms close to natural habitats do not need the industrial-strength pollination services of honey bees. If you eat organic food, then you are encouraging organic agriculture, which is a bee-friendly way of growing food. Even if you do not have a garden, you can buy food that is more pollinator-friendly. You can help the bees in this way even if you live in a high-rise without a balcony.

5. WALK ON THE GRASS

Few ground-nesting bee species like nesting in a dense swath of grass; most prefer the bare spaces created by those who ignore instructions not to walk on the grass. When large numbers of people are allowed to tread along the same shortcut across a lawn, the path may erode. If this creates a narrow slope on either side of the informal footpath, even more bees will be able to nest there, and perhaps some bee-friendly "weeds" will grow. Unfortunately, many lawn owners will "pave" over bare ground with sod. As long as they do this at a time of year when the bees are not active, and they do not dig too deeply as they replace the soil, any bees nesting underneath may be able to survive (though they will likely have to find somewhere else to nest the following spring or summer).

6. ENCOURAGE BEE-FRIENDLY PRACTICES AT VARIOUS GOVERNMENTAL LEVELS

There are numerous things that local, national and international governments can do to help bees. The principles discussed above for private gardens, if applied at a city-wide level, will make a better environment for bees on a larger scale. Encouraging municipalities to seed roadside verges with mixes of native wildflowers will both encourage native pollinators and discourage invasive weed species.

Whenever you get the chance, please support by-laws that promote pesticide-free gardening, bee-friendly urban planning (such as green roofs) and golf course designs, native plant seeding and so on.

New housing developments, high-rises and office towers can benefit from having green roofs. Switching from a standard roof to one that has soil and vegetation on it can result in substantial energy savings. Less heating is required in winter because of the insulation that the soil and vegetation provide. Less cooling is needed in summer because of the heat loss that occurs when water evaporates from the soil. What gets planted on these roofs is important in determining whether bees can live there or not. At my university there are a couple of buildings with green roofs, and some of my students have surveyed them and compared the numbers of bee species found there with the numbers found in the surrounding areas. The green roofs had almost as many bee species as any of the floristically diverse fields that were adjacent to them: almost fifty bee species were found in over two years, and one of them I had never seen before. Not surprisingly, many of the most common species on the roofs were ground-nesting social bees—those that can quickly build up a large population.

∾

So you can help the bees by buying organic, walking on the grass, avoiding mulch, staking plants with bamboo canes, growing raspberries and other pithy-stemmed plants and leaving the canes in place year after year or gathering them together in bundles. You can also grow simpler flowers, drill holes in the walls of your home and any wooden structures on your property, and encourage municipal, national and international governments to promote bee-friendly practices. There's no need to take all these steps at once, but if you can do one or more, you will be contributing to a more sustainable, more nutritious world where the visual and olfactory stimuli will be more diverse. You may not have time to stop and smell the wild roses—or watch the wild bees that are pollinating them—but you make it more likely that future generations will be able to.

Epilogue

Anyone who runs a store, or a business of any sort, has to be certain of the inventory on hand. Any store owner or businessperson who does not have a good idea of that inventory, at least at several "stocktaking" times each year, will soon need to seek a different line of work.

As the dominant species on the planet, we should consider ourselves the storekeeper or factory owner. But do we know what our inventory is? The answer is a resounding no that will echo down through the ages. How much longer we survive depends on our being able to get a good idea of the details of this inventory—that is, the natural capital that sustains the planet. Accounting for this diversity is the task of taxonomists, who are themselves almost a dying breed. Our survival will also depend on our gaining a far better understanding of the interactions among the millions of species that have coexisted with us since we came down from the trees. This is the task of ecologists.

Ecologists want to know how natural systems function. What goes on in a pristine ecosystem has been their purview since the origin of this branch of biology. But given the global impact of human beings, it seems that there is no longer any such thing as a pristine ecosystem. The impenetrable jungles and unfathomed ocean depths remain. But the temperatures in the jungles and the strengths of the ocean currents are being altered by our activities. We badly need to understand our impact on our planet's biological diversity. The first step in this is cataloguing the world's biological components. It is essential that we increase the taxonomic and ecological understanding of our natural capital.

Urban existence permits increasing numbers of people to live in condominiums, shop in subterranean malls, travel by subway and work in large office complexes. Exposure to natural sunlight is optional with this kind of lifestyle. It is possible to go for days, weeks, months and even years seeing only sunlight filtered through windows. The annual cycles of the seasons, the monthly waxing and waning of the moon, the hourly vagaries of the weather—all these things that our ancestors were acutely aware of can pass us by entirely unnoticed. This lack of exposure to the natural world has been referred to as "nature deficit disorder." It encourages the mistaken belief that the non-human world is entirely unnecessary.

How wrong can we be? Until all our food is constructed from chemicals, the air we breathe is generated by machine and the water we drink made from the combustion of hydrogen, we will be absolutely reliant on other species and the ecological systems they contribute to. Our morning mug of coffee relies on pollination provided by a wide range of species. The beef in your hamburger would be enormously more expensive without the activities of the alfalfa leafcutter bees, which pollinate the flowers that produce the seeds that grow into winter forage for cattle. The apple a day that keeps the doctor away would not be available if it weren't for the activities of orchard bees and other species.

You can argue that we have no need to worry because the bulk of our diet comes from wind-pollinated grains and crops pollinated by honey bees. And indeed there is no need to worry if you are happy with a diet of rice, corn, wheat, barley, millet, sorghum and rye. You could enliven the colours of these otherwise monochromatic meals with some artificial dyes (although without pollinators, saffron and turmeric would not be available for long). You could supplement this bland diet with fruits and vegetables brought to you by honey bees, as long as honey bees persist. But even this is becoming an issue.

When I started writing this book, it seemed that the cause of colony collapse disorder had been discovered—a disease. Now things seem less clear. At present, it is possible for North Americans to obtain additional stocks of honey bees by purchasing them from other continents. Australia has been producing more honey bee colonies than it needs, and almond growers (among others) have been supplementing their use of transported homegrown hives with Australian stock. Right now, Australia is free of some of the enemies of honey bees that have depleted hives elsewhere and made the lives of beekeepers more difficult. Varroa mites are not yet found in Australia, although they arrived in New Zealand in 2000 (and so their appearance in Australia is almost guaranteed). When this source of enemy-free honey bee colonies dries up, much of our agriculture will be in trouble.

Pesticides have been implicated in at least some of the declines of honey bee populations, creating an antagonistic relationship between one branch of our food supply system and another. In any complex system, it is destructive to have one part actively negate the effects of another. But this is exactly what seems to be happening in agriculture. The part of the operation that aims to ensure that pests don't consume the crops is killing off the pollinators that make the crops possible in the first place. If you were running a car plant and the folks who painted the vehicles went around slashing the tires, you would be out of business in no time. Agriculture is not so simple. Large numbers of species are essential for the production of most crops. Interactions, both biotic and abiotic, are complex and poorly understood. We need an early warning system that enables us to detect whether a new practice is causing undesirable end results. We need an early warning system that will inform us when some new combination of old practices is causing undesirable effects.

With their essential ecological roles, large numbers of species throughout the globe and high risk of extinction, bees are among the best-available monitors of the state of the world's terrestrial ecosystems. Bee monitoring may become an essential part of our adaptation to our changing global circumstances. We need to be able to assess the state of the environment and use sensitive indicators to do so. I believe that bees are an excellent choice for this work.

We need to get going with this enterprise immediately. Certainly there are some encouraging movements in this direction. For example, recent proclamations by the Convention on Biological Diversity have exhorted member nations to promote pollinator-friendly agricultural practices, and to pay attention to the diversity of their pollinators and work to reverse their declines. Hardly a day goes by without some newspaper running an article about the plight of the bees. The development of user-friendly identification guides and the advent of semi-automated genetic identification tools will also increase the general popularity of bees and other pollinators. But the situation is urgent. Every time a bee species goes extinct, our information-gathering capacity is diminished. Every time a bee species goes extinct, our capacity to develop new pollinators decreases. Every time a bee species goes extinct, the jigsaw puzzle of life loses another important piece. Every time a bee species goes extinct, the world becomes a simpler place—and one that is less suitable for our continued existence.

Five mass extinctions have occurred over the past six hundred million years, and we are causing the sixth. After each of the previous mass extinctions, biodiversity recovered. But the recovery took ten million years in most cases, and thirty million years in the worst one. I do not know about you, but I am not a very patient person; sometimes waiting a few minutes is more than I would like. I doubt if any of us are willing to wait millions of years for biodiversity to recover from our actions.

While many of the advances made by modern civilization have been beneficial, they have come at a cost: we have a lack of regard for the non-human world, and for an increasing majority, this includes a lack of respect for non-urban human communities. Many indigenous cultures are themselves in danger of extinction, but we can learn some approaches to life from them.

In early October 2008, I attended a meeting of Mexican and Central American bee researchers in Merida, Mexico. An evening outing had been arranged by the conference organizer, Javier Quezada-Euan, an up-and-coming researcher with a speciality in stingless bees. He invited us to see the ancient Mayan "feeding the bees" ceremony. The bees involved in the ceremony were stingless bees responsible for pollinating local crops and providing beekeepers with medicinally and nutritionally exquisite honey.

The shaman-like figures started their ritual just before sunset, aided by the burning of incense, the drinking of fermented stingless bee honey and the sounding of conch horns. Apart from a few introductory remarks in Spanish, the ceremony was performed in a Mayan language. These incantations have been handed down from generation to generation through oral tradition for thousands of years. The ancient Mayans worshipped bees, and their descendants continue this practice. Perhaps we should take it up as well.

I hope I have raised your consciousness about bees and their role in our lives. I hope that when you stop and smell the roses, you will now tarry a little longer and look at the bee that is pollinating the flowers. I hope that you will take the time to find out a little more about these beautiful creatures on which so much of our lives depends. Perhaps you will provide homes for bees in your garden or on your apartment balcony. Perhaps you will become absorbed by the complex antics of these newly recognized partners in our passage through life. Perhaps, as my parents did, you will instil in your children a fascination for these unsung heroes and heroines of the natural world.

I hope so. Your grandchildren may depend upon it.

Acknowledgments

It's impossible to acknowledge all the people who have helped me during the more than thirty years that I have been studying bees. Many of those who administer the research grants and help purchase the equipment or manufacture it in the first place are unknown to me—but I thank you all nonetheless. Certainly, as the dedication at the beginning of this book suggests, my parents gave me the initial encouragement, for which I will be forever grateful. David Rogers was my entomology tutor while I was an undergraduate, and Peter Kirby and Donald Quicke were enthusiastic company on collecting trips at that time. George Else assisted me in my early studies of bees at the National History Museum in London (to which Colin Vardy introduced me), and the late John Felton helped identify the specimens in my earliest bee collections. Gerd Knerer took me on as a Ph.D. student, and during those years I was greatly encouraged by John Wenzel and Bill Wcislo and mentored by the late George Eickwort and Christopher Plowright. Charles Michener continues to be an inspiration. My postdoctoral research was conducted under the supervision of Robin Owen.

Since arriving at York University, I have benefited considerably from the advice of numerous colleagues, in particular Marla Sokolowski, Joel Shore, Bridget Stutchbury and Gene Morton, and the forbearance of a series of departmental chairs: Brock Fenton, Brian Colman, Art Hilliker and Imogen Coe. As a university professor, I have been lucky enough to have had a series of graduate students who have taken bee research in new directions and much further than I ever could have done; space constraints permitted me to mention the work of only a fraction of them in this book, but the rest know who they are, and I hope to embarrass more of you in print in the future. My colleagues at the bee course in Arizona (John Ascher, Steve Buchmann, Jim Cane, Bryan Danforth, Terry Griswold, Gretchen LeBuhn, Ron McGinley, Robbin Thorp and especially Jerry Rozen) provide me with a regular topping-up of enthusiasm (and I apologize for annoying them in Chapter 4). The members of the "traiters" working group, Elizabeth Crone, Robert Minckley, Simon Potts, T'ai Roulston and Neal Williams, provided information that was an essential component of Chapter 11.

There are numerous other researchers who have let me come along on field trips they've organized and take part in their research; in addition to some of those mentioned above, I am particularly grateful to Eduardo Almeida, Ricardo Ayala, Connal Eardley, Mary Gikungu, John Heraty, Peter Kevan, Jack Neff, Cecile Plateaux-Quénu, Luc Plateaux, Luisa Ruz, Michael Schwarz, Michael Sharkey and Ken Walker. Even more people have assisted my students in their field research. Our work on barcoding the bees of the world would have been impossible without the assistance of dozens of collectors, museum curators, researchers and other enthusiasts literally throughout the world, as well as the tireless efforts of Paul Hebert, Robert Hanner and

Sujeevan Ratnasingham, among others, at the Biodiversity Institute of Ontario. At various stages I have benefited from the skepticism of my daughters, Rosie and Laurel, the typing prowess of Chloe and the distractions provided by Miko.

My research has been funded primarily by the Natural Sciences and Engineering Research Council of Canada, with some additional assistance from the National Geographic Society, the York University Faculty Association, the Food and Agriculture Organization of the United Nations, the Canadian Fund for Innovation, the Ontario Research Foundation and Genome Canada, among others.

I thank the folks at HarperCollins for their assistance with the various phases in the generation of this book: Jim Gifford, Noelle Zitzer, Janice Weaver. Just about everyone mentioned by name in the book read the portion of the text that pertained to their work, and I am grateful for their corrections, suggestions and support. My labmates Jennifer Albert, Lincoln Best, Sheila Colla, Sheila Dumesh, Jason Gibbs, Cory Sheffield and Alana Taylor each provided helpful comments on some of the chapters. Steve Buchmann, Wayne Goertzen and Steven Morgan-Jones provided timely replies to questions, as again did most of the people mentioned by name in the text. Nonetheless, any errors that remain have to be blamed on me.

I extend a hearty personal thanks to Nancy Davis, Karen Kaffko, Clement Kent, Bill and Fenella Nicholson, Juliet Palmer, Yvonne Pamula, Leena Raudvee, James Rolfe, Marla and Allen Sokolowski, Jonathan Wong and Ruth Wyman.

Finally, I thank my wife, part-time editor, muse, lover and occasional field assistant, Gail Fraser. This book could not have been brought to fruition without her continued support.

Appendix 1: Bee Families

Most melittologists agree that there are seven families of bees. They are separated mostly on the basis of difficult-to-observe characteristics of the mouthparts. Presumably, as soon as bees evolved, the increased need to bring nectar to add to the pollen provided for the young resulted in adaptations to make nectar gathering more efficient. The seven different families are listed here in alphabetical order, along with some of the common names of the bee species that belong to them.

Andrenidae: The solitary mining bees and their relatives. There are thousands of species.

Apidae: An enormous and diverse family that includes nomad, squash, carpenter, orchid, bumble, stingless and honey bees.

Colletidae: The cellophane and masked bees that line their nests with transparent cellophane-like material.

Halictidae: The sweat bees and pearly-banded bees (including the alkali bee). There are several thousand known species.

Megachilidae: The leafcutter, resin, orchard and mason bees. There are several thousand known species.

Melittidae: The oil-collecting bees and others. There are fewer than 250 species.

Stenotritidae: There is no common name for these rare bees. There are fewer than thirty species, all of which are restricted to Australia.

Appendix 2: Bee Names

Throughout this book, I have used the common names for bees rather than the scientific ones. Most of these common names I concocted myself because few bees are well enough known, often enough noticed or sufficiently easily distinguished from related species for people to have named them informally. But if you want to find out more about particular species by surfing the web for images or details on their biology, you'll need to use their scientific names and so these are listed below, alongside the common names used in the text. I also note which bee family each species belongs to and approximately where in the world it may be found. Asterisks denote common names that are actually in use.

COMMON NAME	SCIENTIFIC NAME	FAMILY	RANGE
Alfalfa leafcutter bee*	*Megachile rotundata*	Megachilidae	Temperate regions worldwide
Allodapine bee	Many different genera	Apidae	Tropical Africa, Asia, Australia
Articulated nomad bee	*Nomada articulata*	Apidae	North America
Ashton's cuckoo bumble bee	*Bombus ashtoni*	Apidae	North America (may be extinct)
Banded sweat bee	*Lasioglossum zonulum*	Halictidae	Europe; introduced into North America

COMMON NAME	SCIENTIFIC NAME	FAMILY	RANGE
Band-footed sweat bee	*Lasioglossum cinctipes*	Halictidae	Eastern North America
Bicoloured agapostemon	*Agapostemon virescens*	Halictidae	North America
Calceate sweat bee	*Lasioglossum calceatum*	Halictidae	United Kingdom to Siberia and Japan
Dawson's burrowing bee	*Amegilla dawsoni*	Apidae	Western Australia
Dwarf honey bee*	*Apis florea* (and a couple of others)	Apidae	Arabia, tropical Asia
Franklin's bumble bee*	*Bombus franklini*	Apidae	California, Oregon (may be extinct)
Giant honey bee*	*Apis dorsata*	Apidae	Tropical Asia
Giant resin bee*	*Megachile sculpturalis*	Megachilidae	East Asia; introduced into eastern North America
Golden augochlorella	*Augochlorella aurata*	Halictidae	Eastern and central North America
Golden-haired exomalopsis	*Exomalopsis aureopilosa*	Apidae	Tropical South America, Caribbean
Golden-tailed lithurgus	*Lithurgus chrysurus*	Megachilidae	Western Mediterranean; introduced into northeastern United States
Greyish colletes	*Colletes seminitidus*	Colletidae	Argentina and Chile
Hoary agapostemon	*Agapostemon sericeus*	Halictidae	Eastern and central North America
Hoary squash bee*	*Peponapis pruinosa*	Apidae	Canada, United States, Mexico

COMMON NAME	SCIENTIFIC NAME	FAMILY	RANGE
Honey bee. See Western domesticated honey bee			
Jewelweed bumble bee	*Bombus impatiens*	Apidae	North America
Large carpenter bees	*Xylocopa* (many species)	Apidae	Worldwide
Ligated gregarious bee	*Halictus ligatus*	Halictidae	North America, except the extreme southeast
Long-horned sphecodes	*Sphecodes monilicornis*	Halictidae	Europe, Morocco, Pakistan, India
Macropis cuckoo bee	*Epeoloides pilosula*	Apidae	Eastern North America (almost extinct)
Masked bee	*Hylaeus* (many species)	Colletidae	Worldwide
Messenger sweat bee	*Lasioglossum malachurus*	Halictidae	Europe, Middle East, North Africa
Orchard bee	*Osmia* (many species)	Megachilidae	Northern hemisphere
Pale-footed habropoda	*Habropoda pallipes*	Apidae	Southwestern United States
Pearly-banded bee	*Nomia* (many species)	Halictidae	Worldwide, except Central and South America and the Far North
Pluto resin bee	*Megachile pluto*	Megachilidae	Indonesia
Poey's gregarious bee	*Halictus poeyi*	Halictidae	Southeastern United States, Caribbean
Portal macrotera	*Macrotera portalis*	Andrenidae	Arizona, Mexico, New Mexico

COMMON NAME	SCIENTIFIC NAME	FAMILY	RANGE
Red-bellied leioproctus	*Leioproctus rufiventris*	Colletidae	Chile
Red-tipped anthophora	*Anthophora terminalis*	Apidae	Europe, North America
Red osmia*	*Osmia rufa*	Megachilidae	Europe, North Africa, Middle East, Mongolia
Red southern oxaea	*Notoxaea ferruginea*	Andrenidae	Argentina, Brazil, Paraguay
Rozen's xeromelissa	*Xeromelissa rozeni*	Colletidae	Northern Chile
Ruddy centris	*Centris* sp.	Apidae	Argentina
Ruddy driveway bee	*Halictus rubicundus*	Halictidae	Northern hemisphere
Rusty-patched bumble bee*	*Bombus affinis*	Apidae	Eastern North America (almost impossible to find now)
Small broad-tongued bee from Leyburn	*Euryglossina leyburnensis*	Colletidae	Eastern Australia
Small shiny sweat bee	*Lasioglossum laevissimum*	Halictidae	North America
Three-toothed exoneurella	*Exoneurella tridentata*	Apidae	Western and southern Australia
Virginia carpenter bee*	*Xylocopa virginica*	Apidae	Eastern North America
Vulture bees	Three species in the genus *Trigona*	Apidae	Central America

COMMON NAME	SCIENTIFIC NAME	FAMILY	RANGE
Western bumble bee*	*Bombus occidentalis*	Apidae	Western North America (now very rare)
Western domesticated honey bee (or just honey bee)	*Apis mellifera*	Apidae	Almost worldwide
White-banded sweat bee	*Lasioglossum leucozonium*	Halictidae	United Kingdom to Siberia, North Africa, Middle East; introduced into eastern North America
Wool carder bee*	*Anthidium manicatum*	Megachilidae	Europe, North Africa; introduced into South America, New Zealand, North America
Yellow-banded bumble bee*	*Bombus terricola*	Apidae	Northern North America
Yellow-footed mining bee	*Andrena flavipes*	Andrenidae	United Kingdom to China, North Africa
Zephyr sweat bee	*Lasioglossum zephyrum*	Halictidae	North America

Sources

In the chapter-by-chapter sections below, I begin by listing sources that lay readers may find particularly useful and approachable. Sometimes these are websites or even links to television programs. Following that, I provide some of the reference material on which my text is based. Often there are many similar articles in support of a particular point; I generally chose the article I was most familiar with, although sometimes a paper is listed because it was an early classic. Whenever websites are mentioned, they were generally accessed late in 2009 or in the first few weeks of 2010.

Parents and conservation biologists, and particularly people who are both, will want children to be able to understand the importance of bees and the issues raised by the current extinction crisis. Fortunately, there are several good children's books on these topics. Some of these are available through the Pollinator Partnership (*www.pollinator.org/store.htm*). *The Bee Tree* by Stephen Buchmann, Diana Cohn and Paul Mirocha is well worth a special mention. There are also some resources available for teachers wanting to bring pollination information into their classrooms.

See *www.seeds.ca/proj/poll/index.php?n=For+Kids+and+Teachers!* for an introduction. Finally, for those troublesome teenage years when comics are the only acceptable form of reading, try Jay Hosler's Clan Apis (*www.jayhosler.com/clanapis.html*). It could very well result in your adolescent becoming a beekeeper.

1. BUZZ FREE: A WORLD WITHOUT BEES

In recent years, numerous scholarly articles and quite a few popular books have described the economic importance of bees and the impact of the terrible (and terrifying) losses of honey bees. The books are mostly written by non-scientists and generally provide a somewhat incomplete picture. A good summary of our current understanding of the decline in honey bees can be found in a *Nature of Things* documentary, available (in Canada only, alas) at *www.cbc.ca/video/#/Shows/The_Nature_of_Things/ID=1380312270*. The book *Status of Pollinators in North America,* by the Committee on the Status of Pollinators in North America, is a comprehensive survey that covers the period immediately before the rise of colony collapse disorder. The article by Losey and Vaughan is a good overview of the wide range of economically important things insects do for us, and the book by Stephen Buchmann and Gary Nabhan is a good overall introduction to pollinator conservation.

Aizen, Marcelo A., and Lawrence D. Harder. "The Global Stock of Domesticated Honey Bees Is Growing Slower Than Agricultural Demand for Pollination." *Current Biology* 19 (2009): 1–4.

Buchmann, Stephen L., and Gary Paul Nabhan. *The Forgotten Pollinators.* Washington, DC: Island Press, 1995.

Committee on the Status of Pollinators in North America. *Status of Pollinators in North America.* Washington, DC: National

Research Council, 2007.

Gallai, Nicola, Jean-Michel Salles, Josef Settele and Bernard E. Vaissière. "Economic Valuation of the Vulnerability of World Agriculture Confronted with Pollinator Decline." *Ecological Economics* 68 (2009): 810–21.

Hartley, Adrian J., Guillermo Chong, John Houston and Anne E. Mather. "150 Million Years of Climatic Stability: Evidence from the Atacama Desert, Northern Chile." *Journal of the Geological Society* 162 (2005): 421–24.

Kasina, John M., J. Mburu, Manfred Kraemer, K. Holm-Mueller. "Economic Benefit of Crop Pollination by Bees: A Case of Kakamega Small-holder Farming in Western Kenya." *Journal of Economic Entomology* 102 (2009): 467–73.

Klein, Alexandra-Maria, Bernard E. Vaissière, James H. Cane, Ingolf Steffan-Dewenter, Saul A. Cunningham, Claire Kremen and Teja Tscharntke. "Importance of Pollinators in Changing Landscapes for World Crops." *Proceedings of the Royal Society Series B* 274 (2007): 303–13.

Losey, John E., and Mace Vaughan. "The Economic Value of Ecological Services Provided by Insects." *BioScience* 56 (2006): 311–23.

Partap, Uma, T. E. J. Partap and H. E. Yonghua. "Pollination Failure in Apple Crop and Farmers' Management Strategies in Hengduan Mountains, China." Paper presented at the Eighth International Symposium on Pollination, Mosonmagyaróvár, Hungary.

Schneider, Stanley S., Gloria DeGrandi-Hoffman and Deborah R. Smith. "The African Honey Bee: Factors Contributing to a Successful Biological Invasion." *Annual Review of Entomology* 49 (2004): 351–76.

Watanabe, Myrna E. "Colony Collapse Disorder: Many Suspects, No Smoking Gun." *BioScience* 58 (2008): 384–88.

2. THE FUTURE OF OUR FOOD

A good overview of the importance of biodiversity for agriculture is *Managing Biodiversity in Agricultural Ecosystems,* edited by Devra I. Jarvis, Christine Padoch and H. David Cooper (New York: Columbia University Press, 2007). The chapter in that book by Peter G. Kevan and Victoria A. Wojcik covers pollinator biodiversity, and the one by Timothy Johns discusses the importance of dietary diversity for human health. The article listed below by Klein and colleagues provides a detailed account of the role of pollinators on a crop-by-crop basis. A detailed guide on pollination biology methodologies is *Practical Pollination Biology* by Amots Dafni, Peter G. Kevan and Brian C. Husband (Guelph, ON: Enviroquest Ltd., 2008). Details on alfalfa can be found at *http://gears. tucson.ars.ag.gov/book/chap_1.html*

Javorek, Steven K., Kenna E. Mackenzie and Sam P. Vander Kloet. "Comparative Pollination Effectiveness Among Bees (Hymenoptera: Apoidea) on Lowbush Blueberry (Ericaceae: *Vaccinium angustifolium*)." *Annals of the Entomological Society of America* 95 (2002): 345–51.

Kemp, William P., and Jordi Bosch. "Development and Emergence of the Alfalfa Pollinator *Megachile rotundata* (Hymenoptera: Megachilidae)." *Annals of the Entomological Society of America* 93 (2000): 904–11.

Klein, Alexandra-Maria, Bernard E. Vaissière, James H. Cane, Ingolf Steffan-Dewenter, Saul A. Cunningham, Claire Kremen and Teja Tscharntke. "Importance of Pollinators in Changing Landscapes for World Crops." *Proceedings of the Royal Society Series B* 274 (2007): 303–13.

Kremen, Claire, Neal M. Williams and Robbin W. Thorp. "Crop Pollination from Native Bees at Risk from Agricultural Intensification." *Proceedings of the National Academy of Sciences of the United States of America* 99 (2002): 16812–16.

Larsen, Trond H., Neal M. Williams and Claire Kremen. "Extinction Order and Altered Community Structure Rapidly Disrupt Ecosystem Functioning." *Ecology Letters* 8 (2005): 538–47.

Parker, Frank D., Suzanne W. T. Batra and Vincent J. Tepedino. "New Pollinators for Our Crops." *Agricultural Zoology Reviews* 2 (1987): 279–304.

Ricketts, Taylor H., Gretchen C. Daily, Paul R. Ehrlich and Charles D. Michener. "Economic Value of Tropical Forest to Coffee Production." *Proceedings of the National Academy of Sciences of the United States of America* 101 (2004): 12579–82.

Roubik, David W. "Tropical Agriculture: The Value of Bees to the Coffee Harvest." *Nature* 417 (2002): 708.

Sampson, Blair J., and James H. Cane. "Pollination Efficiencies of Three Bee (Hymenoptera: Apoidea) Species Visiting Rabbiteye Blueberry." *Journal of Economic Entomology* 93 (2000): 1726–31.

3. HONEY, QUEENS, HARD-WORKING WORKERS AND STINGS: MISCONCEPTIONS ABOUT BEES

The Bees of the World by Charles Michener is a thorough compendium of just about everything known about bees up until 2006. His *Social Behavior of the Bees* provides detailed information about the social lives of bees, but it is now somewhat out of date. Chris O'Toole and Anthony Raw's book is an easy read on the basics of bee biology.

Michener, Charles D. *The Bees of the World.* Baltimore: John Hopkins University Press, 2007.

Michener, Charles D. *The Social Behavior of the Bees.* Cambridge, MA: Belknap Press, 1974.

O'Toole, Christopher, and Anthony Raw. *Bees of the World.* London: Blandford Press, 2004.

Packer, Laurence. "The Comparative Morphology of the Skeletal Parts of the Sting Apparatus of Bees (Hymenoptera: Apoidea)." *Zoological Journal of the Linnean Society* 138 (2003): 1–38.

Raffiudin, Rika, and Ross H. Crozier. "Phylogenetic Analysis of Honey Bee Behavioral Evolution." *Molecular Phylogenetics and Evolution* 43 (2007): 543–52.

Schmidt, Justin O. "Hymenoptera Venoms: Striving Toward the Ultimate Defense Against Vertebrates." In *Insect Defenses: Adaptive Mechanisms and Strategies of Prey and Predators.* Edited by D. L. Evans and Justin O. Schmidt. Albany: State University of New York Press, 1990.

4. A BEE OR NOT A BEE? A DIFFICULT QUESTION TO ANSWER
Judith Winston's *Describing Species* outlines practical procedures for taxonomists but is aimed at an audience with training in biology. A very readable account for the lay person is Carol Kaesuk Yoon's *Naming Nature: The Clash Between Instinct and Science* (New York: Norton, 2009). The clash in the title refers to the differences between classifications based upon overall appearance (the instinct) and those that use evolutionary trees (the science). Some

of the more general items below discuss what bees are. The key by Packer and Ratti provides images of the main characteristics. Many websites have images of bees and related insects, but too many have non-bees identified as bees and bees identified as something else. An excellent introduction to bee classification and evolution can be found on Bryan Danforth's website at *www.danforthlab.entomology.cornell.edu/content/view/bee-diversity.html*.

Engel, Michael S. "A Monograph of the Baltic Amber Bees and Evolution of the Apoidea (Hymenoptera)." *Bulletin of the American Museum of Natural History* 259 (2001): 1–192.

Goulet, Henri, and John T. Huber. *Hymenoptera of the World: An Identification Guide to Families.* Ottawa: Agriculture Canada, 1993.

O'Toole, Christopher, and Anthony Raw. *Bees of the World.* London: Blandford Press, 2004.

Packer, Laurence, and Claudia M. Ratti. *Key to the Bee Families of the World.* Accessed online at *www.yorku.ca/bugsrus/BFoW/Images/Introduction/Introduction.html*.

Plowright, R. Christopher, and Robin E. Owen. "The Evolutionary Significance of Bumble Bee Colour Patterns: A Mimetic Interpretation." *Evolution* 34 (1980): 622–37.

Poinar, George O., and Bryan N. Danforth. "A Fossil Bee from Early Cretaceous Burmese Amber." *Science* 314 (2006): 614.

Ruxton, Graeme D., Thomas N. Sherratt and Michael P. Speed. *Avoiding Attack: The Evolutionary Ecology of Crypsis, Warning Signals and Mimicry.* Oxford: Oxford University Press, 2004.

Sharkey, Michael J. "Phylogeny and Classification of the Hymenoptera." *Zootaxa* 1668 (2007): 521–48.

Winston, Judith E. *Describing Species: Practical Taxonomic Procedure for Biologists.* New York: Columbia University Press, 1999.

5. TWO BEES OR NOT TWO BEES? AN EVEN *MORE* DIFFICULT QUESTION TO ANSWER

If you want to learn how to collect and identify bees, there are several courses you can take. For information on the bee course mentioned in this book, consult the website at *research.amnh.org/iz/beecourse/*. Another website that provides a good overview can be found at *www.nbii.gov/images/uploaded/152986_1244054830561_ Handy_Bee_Manual_Jun_2009.pdf.* The Canadian Journal of Arthropod Identification provides numerous online, beautifully illustrated keys to Canadian insects, including a key to the bee genera of eastern Canada. A key to the bee genera of the whole country will be added soon. Additional identification guides to the bees of Canada will appear in other online journals, including *Zootaxa* (*www.mapress.com/zootaxa/*) and *The Canadian Entomologist* (*www.esc-sec.ca/journal.html*), over the next few years. For guides to the bees of all of North America, consult the Discover Life website at *www.discoverlife.org/20/q?search=Apoidea.* On that site you will also find lists of all bee species found in the world, as well as country-by-country lists and maps showing the distributions of all species. *The Bee Genera of North and Central America,* listed below, is a useful guide, although the classification system used is somewhat dated (and because it is out of print, the book now fetches prices that make me wish I had bought one hundred copies fifteen years ago). *The Bees of the World* provides references to identification guides at the species level for all bees for which such keys have been written. The current status of the bee barcoding campaign can be found at *www.bee-bol.org* and on the

Barcode of Life website at *www.barcodinglife.org/views/login.php*.
Large numbers of images of bees can also be found on these websites, as well as on BugGuide (*bugguide.net/node/view/15740*). At
this latter site, people can post pictures of any bug and solicit the
assistance of others to identify it.

Donovan, Barry J. *Apoidea (Insecta: Nymenoptera)*. Fauna of New
    Zealand, Ko te Aitanga Pepeke o Aotearoa. Number 57. Mana-
    oki Whenua Press.

Hebert, Paul D. N., Erin H. Penton, John M. Burns, Daniel H.
    Janzen and Winnie Hallwachs. "Ten Species in One: DNA
    Barcoding Reveals Cryptic Species in the Neotropical Skipper
    Butterfly *Astraptes fulgerator*." *Proceedings of the National Academy
    of Sciences of the United States of America* 101 (2004): 14812–17.

International Commission on Zoological Nomenclature. *Interna-
    tional Code of Zoological Nomenclature*. London: International
    Trust for Zoological Nomenclature, 1999.

Michener, Charles D. *The Bees of the World*. Baltimore: John
    Hopkins University Press, 2007.

Michener, Charles D., Ronald J. McGinley and Bryan N. Dan-
    forth. *The Bee Genera of North and Central America* (Hymenop-
    tera: Apoidea). Washington, DC: Smithsonian Institution Press,
    1994.

Packer, Laurence, Jason Gibbs, Cory Sheffield and Robert
    Hanner. "DNA Barcoding and the Mediocrity of Morphology."
    Molecular Ecology Resources 9 (2009): 42–50.

Winston, Judith E. *Describing Species: Practical Taxonomic Proce-
    dure for Biologists*. New York: Columbia University Press, 1999.

The O'Toole and Raw book provides a user-friendly overview of bee biology. Michael Schwarz's website (*www.scieng.flinders. edu.au/current/biology/mps/mpsbees1.htm*) contains a lot of useful information on bees, as well as some beautiful images, including several of the trident-tailed exoneurella discussed at the beginning of this chapter.

Almeida, Eduardo A. B. "Colletidae Nesting Biology (Hymenoptera: Colletidae)." *Apidologie* 39 (2008): 16–29.

Bänziger, Hans, Somnuk Boongird, Prachaval Sukumalanand and Sängdao Bänziger. "Bees (Hymenoptera: Apidae) That Drink Human Tears." *Journal of the Kansas Entomological Society* 82 (2009): 135–50.

Camargo, João M. F., and David W. Roubik. "Systematics and Bionomics of the Apoid Obligate Necrophages: The *Trigona hypogeal* Group." *Biological Journal of the Linnean Society* 44 (1991): 13–39.

Eickwort, George C., and Jerome G. Rozen. "The Entomological Evidence." *Journal of Forensic Science* 42 (1997): 394–97.

Mateus, Sidnei, and Fernando B. Noll. "Predatory Behavior in a Necrophagous Bee *Trigona hypogea* (Hymenoptera: Apidae, Melipnini)." *Naturwissenschaften* 91 (2004): 94–96.

Michener, Charles D. *The Bees of the World.* Baltimore: John Hopkins University Press, 2007.

Neff, John L. "Components of Nest Provisioning Behavior in Solitary Bees (Hymenoptera: Apoidea)." *Apidologie* 39 (2008): 30–45.

O'Toole, Christopher, and Anthony Raw. *Bees of the World.* London: Blandford Press, 2004.

Packer, Laurence. "Solitary and Eusocial Nests in a Population of *Augochlorella striata* (Provancher) (Hymenoptera: Halictidae) at the Northern Edge of Its Range." *Behavioral Ecology and Sociobiology* 27 (1990): 339–44.

Packer, Laurence, Blair Sampson, Cathy Lockerbie and Vincent Jessome. "Nest Architecture and Brood Mortality in Four Species of Sweat Bee (Hymenoptera: Halictidae) from Cape Breton Island." *Canadian Journal of Zoology* 67 (1989): 2864–70.

Praz, Christophe J., Andreas Müller and Silvia Dorn. "Specialized Bees Fail to Develop on Non-host Pollen: Do Plants Chemically Protect Their Pollen?" *Ecology* 89 (2008): 795–804.

Schwarz, Michael P., Miriam H. Richards and Bryan N. Danforth. "Changing Paradigms in Insect Social Evolution: Insights from Halictine and Allodapine Bees." *Annual Review of Entomology* 52 (2006): 127–50.

Steiner, Kim E., and Vim B. Whitehead. "Pollinator Adaptation to Oil-secreting Flowers—*Rediviva* and *Diascia*." *Evolution* 44 (1990): 1701–07.

7. THE SOCIABLE BEE

Michener's *Social Behavior of the Bees* remains a thorough study of bee social behaviour up until the mid-1970s. A lively account of the role of genetic relatedness in promoting social evolution in bees, wasps and ants can be found in Richard Dawkins's *The Selfish Gene* (Oxford: Oxford University Press, 2006). More detailed studies of social evolution in a wide range of insects can be found in *The Evolution of Social Behavior in Insects and Arachnids*, edited by Jae C. Choe and Bernard J. Crespi (Cambridge, UK: Cambridge University Press, 1997), which has several chapters that deal with bees. Several very readable accounts of honey bee biology have been written. The most highly recommended are *Honeybee Ecology: A Study of Adaptation in Social Life* (Princeton, NJ: Princeton University Press, 1985) and *The Wisdom of the Hive: The Social Physiology of Honey Bee Colonies* (Cambridge, MA: Harvard University Press, 1995), both by Tom Seeley, and *The Biology of the Honey Bee* by Mark Winston (Cambridge, MA: Harvard University Press, 1987). There is even a scientifically accurate comic on honey bees: *Clan Apis* by Jay Hosler (*www.jayhosler.com/clanapis.html*). Stephen Buchmann has written a book about social bees aimed at a teenage audience: *Honeybees: Letters from the Hive* (New York: Delacourt Press, 2010). Michael Schwarz's website (*www.scieng.flinders.edu.au/current/biology/mps/mpsbees1.htm*) contains a lot of useful information on social bees.

Abrams, Judith, and George C. Eickwort. "Nest Switching and Guarding by the Communal Sweat Bee *Agapostemon virescens* (Hymenoptera: Halictidae)." *Insectes Sociaux* 28 (1981): 105–16.

Alexander, Richard D. "The Evolution of Social Behavior." *Annual Review of Ecology and Systematics* 5 (1974): 325–83.

Andersson, Malte. "The Evolution of Eusociality." *Annual Review of Ecology and Systematics* 15 (1984): 165–89.

Duchateau, Marie J., and Hayo H. W. Velthuis. "Development and Reproductive Strategies in *Bombus terrestris* Colonies." *Behaviour* 107 (1988): 186–207.

Hughes, William O., Benjamin P. Oldroyd, Madeleine Beekman and Francis L. W. Ratnieks. "Ancestral Monogamy Shows Kin Selection Is Key to the Evolution of Eusociality." *Science* 320 (2008): 1213–16.

Michener, Charles D. *The Social Behavior of the Bees*. Cambridge, MA: Belknap Press, 1974.

Packer, Laurence. "The Social Organisation of *Lasioglossum (Dialictus) laevissimum* (Smith) in Southern Alberta." *Canadian Journal of Zoology* 70 (1992): 1767–74.

Plateaux-Quénu, Cecile. "Le Volume d'un pain d'abeille influence t'il le sexe de l'oeuf pondu sur lui? Étude expérimentale, portant sur la première couvée d'*evylaeus calceatus* (Scop.)." *Annales des Sciences Naturelles, Zoologie, Paris* 13 (1983): 41–52.

Queller, David, and Joan E. Strassmann. "Kin Selection and Social Insects." *BioScience* 48 (1998): 65–75.

Wilson, Edward O. *Sociobiology: The New Synthesis.* New York: John Wiley, 1975.

8. SEX AND DEATH IN BEES

For an entertaining account of the weird and wonderful world of sex in the rest of the planet's biodiversity, you could consult Olivia Judson's *Dr. Tatiana's Sex Advice to All Creation* (London: Chatto and Windus, 2002), though you may find the style grates after a while. Genetics is a complex subject that can quickly leave a reader flummoxed. Richard Frankham, Jonathan D. Ballou and David A. Briscoe's *A Primer of Conservation Genetics* (Cambridge, UK: Cambridge University Press, 2004) provides a good introduction to the subject as applied to conservation, and Packer and Owen's paper, cited below, is an early summary of the subject as applied to pollinators.

Hedrick, Philip W., Jürgen Gadau and Robert E. Page. "Genetic Sex Determination and Extinction." *Trends in Ecology and Evolution* 21 (2006): 55–57.

Packer, Laurence. "The Biology of a Subtropical Population of *Halictus ligatus* Say (Hymenoptera, Halictidae). II: Male Behaviour." *Ethology* 72 (1986): 287–98.

Packer, Laurence, and Robin E. Owen. "Population Genetic Aspects of Pollinator Decline." *Conservation Ecology* 5 (2001): 4. Accessed online at *www.consecol.org/v0l5/iss1/art4/*.

Starks, Phillip T., and Hudson K. Reeve. "Condition-based Alternative Reproductive Tactics in the Wool-carder Bee *Anthidium manicatum*." *Ethology, Ecology and Evolution* 11 (1999): 71–75.

Zayed, Amro. "Bee Genetics and Conservation." *Apidologie* 40 (2009): 237–62.

Zayed, Amro, and Laurence Packer. "Complementary Sex Determination Substantially Increases Extinction Proneness of Haplodiploid Populations." *Proceedings of the National Academy of Sciences of the United States of America* 102 (2005): 10742–46.

Zayed, Amro, David W. Roubik and Laurence Packer. "Use of Diploid Male Frequency Data as an Indicator of Pollinator Decline." *Proceedings of the Royal Society of London Series B* 271 (2004): S9–S12.

9. WHERE THE BEE SUCKS, THERE HUNT I

There is an online discussion group for bee-monitoring protocols at *tech.groups.yahoo.com/group/beemonitoring/*. Sam Droege is also developing a series of videos on how to do this kind of work; these are available online at *www.youtube.com/user/swdroege*. Almost any biogeography text will include some information on the large-scale patterns in biodiversity discussed in this chapter.

Danforth, Bryan N., Shuqing Ji and Luke J. Ballard. "Gene Flow and Population Structure in an Oligolectic Desert Bee, *Macrotera (Macroteropsis) portalis* (Hymenoptera: Andrenidae)." *Journal of the Kansas Entomological Society* 76 (2003): 221–35.

Droege, Sam, Vincent J. Tepedino, Gretchen LeBuhn, William

Link, Robert L. Minckley, Qian Chen and Casey Conrad. "Spatial Patterns of Bee Captures in North American Bowl-trapping Surveys." *Insect Conservation and Diversity* 3 (2010): 15–23.

Grixti, Jennifer C., and Laurence Packer. "Changes in the Bee Fauna (Hymenoptera: Apoidea) of an Old Field Site in Southern Ontario, Revisited After 34 Years." *The Canadian Entomologist* 138 (2006): 147–64.

Kerr, Jeremy, and Laurence Packer. "Habitat Heterogeneity as a Determinant of Mammal Species Richness in High-energy Regions." *Nature* 385 (1997): 252–54.

Minckley, Robert L. "Faunal Composition and Species Richness Differences of Bees (Hymenoptera: Apiformes) from Two North American Regions." *Apidologie* 39 (2008): 176–88.

Pimentel, David, Rodolfo Zuniga and Doug Morrison. "Update on the Environmental and Economic Costs Associated with Alien-invasive Species in the United States." *Ecological Economics* 52 (2005): 273–88.

Roberts, Radclyffe B. "The Nesting Biology, Behavior and Immature Stages of *Lithurge chrysurus,* an Adventitious Wood-boring Bee in New Jersey (Hymenoptera: Megachilidae)." *Journal of the Kansas Entomological Society* 51 (1978): 735–45.

Westphal, Catrin, Riccardo Bommarco, Gabriel Carré, Ellen Lamborn, Nicolas Morison, Theodora Petanidou, Simon G. Potts, Stuart P. M. Roberts, Hajnalka Szentgyörgyi, Thomas Tscheulin, Bernard E. Vaissière, Michal Woyciechowski, Jacobus C. Biesmeijer, William E. Kunin, Josef Settele and Ingolf Steffan-Dewenter. "Measuring Bee Diversity in Different European Habitats and Biogeographic Areas." *Ecological Monographs* 78 (2008): 653–71.

Zayed, Amro, Laurence Packer, Jennifer C. Grixti, Luisa Ruz,

Robin E. Owen and Haroldo Toro. "Increased Genetic Differentiation in a Specialist Versus a Generalist Bee: Implications for Conservation." *Conservation Genetics* 6 (2005): 1017–26.

10. ANTI-BEES

There are no general overviews of the natural enemies of bees that I am aware of, but the book by Ruxton and colleagues, listed below, surveys defence mechanisms in general, and the one by Schmid-Hempel is a good study of some of the enemies of social bees. Video clips of those evil little beetle larvae transferring to a bee are available online at *www.pnas.org/content/103/38/14039/suppl/DC1#M1*.

Dukas, Reuven. "Bumble Bee Predators Reduce Pollinator Density and Plant Fitness." *Ecology* 86 (2005): 1401–06.

Knerer, Gerd. "Periodizität und Strategie der Schmarotzer einer sozialen Schmalbiene, *Evylaeus malachurus* (K.) (Apoidea: Halictidae)." *Zoologische Anzeiger* 190 (1993): 41–63.

Michener, Charles D. *The Bees of the World*. Baltimore: John Hopkins University Press, 2007.

Norden, Beth B., Karl V. Krombein, Mark A. Deyrup and Jayanthi P. Edirisinghe. "Biology and Behavior of a Seasonally Aquatic Bee, *Perdita (Alloperdita) floridensis* Timberlake (Hymenoptera: Andrenidae: Panurginae)." *Journal of the Kansas Entomological Society* 76 (2003): 236–49.

Ruxton, Graeme D., Thomas N. Sherratt and Michael P. Speed. *Avoiding Attack: The Evolutionary Ecology of Crypsis, Warning Signals and Mimicry*. Oxford: Oxford University Press, 2004.

Saul-Gershenz, Leslie S., and Jocelyn G. Millar. "Phoretic Nest Parasites Use Sexual Deception to Obtain Transport to Their

Host's Nest." *Proceedings of the National Academy of Sciences of the United States of America* 103 (2006): 14039–44.

Schmid-Hempel, Paul. *Parasites in Social Insects.* Princeton, NJ: Princeton University Press, 1999.

11. WHAT ARE WE DOING TO THE BEES?

There are many books that give an overview of extinction. The best general treatment of habitat fragmentation is David Quammen's *Song of the Dodo* (New York: Scribner, 1996). Peter Coates's *American Perceptions of Immigrant and Invasive Species: Strangers on the Land* (Berkeley: University of California Press, 2006) is a lively account of invasive species and their impacts. For pollinators in general, and bees in particular, consult the Xerces Society website (www.xerces.org/) and its red list of endangered bees. *The Forgotten Pollinators* by Stephen Buchmann and Gary Nabhan (Washington, DC: Island Press, 1995), the first book to specifically address the threats faced by pollinators, remains a good read. An entire volume of the journal *Apidologie* ("Bee Conservation," *Apidologie* 40, edited by Robert J. Paxton, Mark J. F. Brown and Thomas E. Murray) was recently devoted to bee conservation. An earlier book on a similar topic is *Bees and Crop Pollination: Crisis, Crossroads and Conservation,* edited by Constance S. Stubbs and Francis A. Drummond (Lanham, MD: Entomological Society of America, 2001).

J. C. Biesmeijer, S. P. Roberts, M. Reemer, R. Ohlemueller, M. Edwards, T. Peeters, A. Schaffers, S. G. Potts, R. Kleukers, C. D. Thomas, J. Settele, and W. E. Kunin. "Parallel Declines in Pollinators and Insect-pollinated Plants in Northwest Europe, Britain and the Netherlands." *Science* 313 (2006): 351–354.

Cane, James H. "Exotic Nonsocial Bees (Hymenoptera: Api-formes) in North America: Ecological Implications." In *For Nonnative Crops, Whence Pollinators of the Future?* Edited by K. Strickler and James H. Cane. Lanham, MD: Entomological Society of America, 2003.

Clavero, M., and E. Garcia-Berthou. "Invasive Species Are a Leading Cause of Animal Extinction." *Trends in Ecology and Evolution* 20 (2005): 110.

Colla, Sheila R., Michael C. Otterstatter, Robert J. Gegear and James D. Thomson. "Plight of the Bumble Bee: Pathogen Spillover from Commercial to Wild Populations." *Biological Conservation* 129 (2006): 461–67.

Colla, Sheila R., and Laurence Packer. "Evidence for Decline in Eastern North American Bumblebees (Hymenoptera: Apidae), with Special Focus on *Bombus affinis* Cresson." *Biodiversity and Conservation* 17 (2008): 1379–91.

Costanza, Robert, Ralph d'Arge, Rudolf de Groot, Stephen Farber, Monica Grasso, Bruce Hannon, Karin Limburg, Shahid Naeem, Robert V. O'Neill, Jose Paruelo, Robert G. Raskin, Paul Sutton and Marjan van den Belt. "The Value of the World's Ecosystem Services and Natural Capital." *Ecological Economics* 25 (1998): 3–15.

Ehrlich, Paul, and Anne Ehrlich. *Extinction: The Causes and Consequences of the Disappearance of Species.* New York: Random House, 1981.

Evans, Elaine, Robbin Thorp, Sarina Jepsen and Scott Hoffman Black. "Status Review of Three Formerly Common Species of Bumble Bee in the Subgenus *Bombus.*" Accessed online at *www.xerces.org/wp-content/uploads/2008/12/xerces_2008_bombus_status_review1.pdf.*

Griswold, Terry, Stacy Higbee and Olivia Messenger. *Pollination Ecology: Final Report.* 2006.

Huntley, Brian, Albert S. van Jaarsveld, Guy F. Midgley, Lera Miles, Miguel A. Ortega-Huerta, A. Townsend Peterson, Oliver L. Phillips and Stephen E. Williams. "Extinction Risk from Climate Change." *Nature* 427 (2004): 145–48.

Kevan, Peter G. "Forest Application of the Insecticide Fenitrothion and Its Effect on Wild Bee Pollinators (Hymenoptera: Apoidea) of Lowbush Blueberries (*Vaccinium* spp.) in Southern New Brunswick, Canada." *Biological Conservation* 7 (1975): 301–09.

May, Robert M. "Thresholds and Breakpoints in Ecosystems with a Multiplicity of Stable States." *Nature* 269 (1977): 471–77.

Pimentel, David, Rodolfo Zuniga and Doug Morrison. "Update on the Environmental and Economic Costs Associated with Alien-invasive Species in the United States." *Ecological Economics* 52 (2005): 273–88.

Stout, Jane C., and Carolina L. Morales. "Ecological Impacts of Invasive Alien Species on Bees." *Apidologie* 40 (2009): 388–409.

Thomas, Chris D., Alison Cameron, Rhys E. Green, Michel Bakkenes, Linda J. Beaumont, Yvonne C. Collingham, Barend F. N. Erasmus, Marinez F. de Siqueira, Alan Grainger, Lee Hannah, Lesley Hughes, Brian Huntley, Albert S. van Jaarsveld, Guy F. Midgley, Lera Miles, Miguel A. Ortega-Huerta, A. Townsend Peterson, Oliver L. Phillips and Stephen E. Williams. "Extinction Risk from Climate Change." *Nature* 427 (2004) 145–48.

Williams, Paul, Sheila Colla and Zhenghua Xie. "Bumblebee Vulnerability: Common Correlates of Winners and Losers on Three Continents." *Conservation Biology* 23 (2009): 931–40.

12. THE PROVERBIAL CANARIES IN THE COAL MINE

There have been several large-scale analyses of how humans rely on the world's biodiversity and the state that biodiversity is in. *Sustaining Life: How Human Health Depends on Biodiversity,* edited by Eric Chivian and Aaron Bernstein (New York: Oxford University Press, 2008), is a recent account. The various volumes issued under the rubric of the Millennium Ecosystem Assessment (*www.millenniumassessment.org/en/index.aspx*) provide large amounts of information from a variety of viewpoints.

Costanza, Robert, Ralph d'Arge, Rudolf de Groot, Stephen
    Farber, Monica Grasso, Bruce Hannon, Karin Limburg, Shahid
    Naeem, Robert V. O'Neill, Jose Paruelo, Robert G. Raskin,
    Paul Sutton and Marjan van den Belt. "The Value of the
    World's Ecosystem Services and Natural Capital." *Ecological
    Economics* 25 (1998): 3–15.
Kevan, Peter G. "Pollinators as Bioindicators of the State of the
    Environment: Species, Activity and Diversity." *Agriculture,
    Ecosystems and Environment* 74 (1999): 373–93.
Nevo, Eviatar, Avigdor Beiles and Rachel Ben-Shlomo. "The
    Evolutionary Significance of Genetic Diversity: Ecological,
    Demographic and Life History Correlates." In *Evolutionary
    Dynamics of Genetic Diversity: Lecture Notes in Biomathemat-
    ics.* Edited by G. S. Mani. Berlin: Springer, 1984.
Packer, Laurence, and Robin E. Owen "Population Genetic
    Aspects of Pollinator Decline." *Conservation Ecology* 5 (2001).
    Accessed online at *www.consecol.org/vol5/iss1/art4*.
Packer, Laurence, Amro Zayed, Jennifer C. Grixti, Luisa Ruz,
    Robin E. Owen, Felipe Vivallo and Haroldo Toro. "Conserva-
    tion Genetics of Potentially Endangered Mutualisms: Reduced

Levels of Genetic Variation in Specialist Versus Generalist
Bees." *Conservation Biology* 19 (2005): 195–202.

Reyes-Novelo, Enrique, Virginia Meléndez-Ramiréz, Hugo
Delfín-González and Ricardo Ayala. "Wild Bees (Hymenop-
tera: Apoidea) as Bioindicators in the Neotropics." *Tropical and
Subtropical Agroecosystems* 10 (2009): 1–13.

13. HELP THE BEES

The website for the David Suzuki Foundation (*www.davidsuzuki.
org*) provides a lot of information on how to improve the planet
in various ways, and not only for bees. For information specific to
helping bees, the Xerces Society website (*www.xerces.org/pollinator-
conservation/*) and that provided by the North American Pollinator
Protection Campaign and the Pollinator Partnership (www.polli-
nator.org) are the best places to start. The book *Befriending Bumble
Bees: A Practical Guide to Raising Local Bumble Bees* by Elaine
Evans, Ian Burns and Marla Spivak (Minneapolis: University of
Minnesota Press, 2007) provides all the information you will need
to be able to attract these beautiful and fascinating insects. Brian
L. Griffin's book *The Orchard Mason Bee* (Bellingham, WA: Knox
Cellars Publishing, 1993) provides useful guidelines that will work
for other species also. There are two helpful booklets available
online from the state of Delaware. One is on farm practices for
wild bees (*dda.delaware.gov/plantind/forms/publications/FarmMan-
agementforNativeBees-AGuideforDelaware.pdf*), and the other is
on planting native wildflowers for bees (*dda.delaware.gov/plantind/
forms/publications/Delaware%20Native%20Plants%20for%20
Native%20Bees.pdf*). In Canada, the David Suzuki Foundation
has published a booklet on wildflowers that will attract pollinators
to your backyard (*www.davidsuzuki.org/files/SWAG/Species/Plant_*

Guide_5pg.pdf) and another on some of the common pollinators (*www.davidsuzuki.org/files/SWAG/Species/Pollinator_Guide_5pg.pdf*). Although the latter specifically covers the Toronto area, it will be useful for almost all of Southern Canada.

Buchmann, Stephen. *Honey for the Maya*. DVD, available from the director: stephenbuchmann@comcast.net.

Colla, Sheila R., Erin Willis and Laurence Packer. "Can Green Roofs Provide Habitat for Urban Bees?" *Cities and the Environment* 2 (2009): 1–12. Accessed online at *escholarship.bc.edu/cgi/viewcontent.cgi?article=1017&context=cate*.

Sheffield, Cory S., Sue M. Westby, Robert F. Smith, and Peter G. Kevan. "Potential of Bigleaf Lupine for Building and Sustaining *Osmia lignaria* Populations for Pollination of Apple." *Canadian Entomologist* 140 (2008): 589–99.

Velthuis, Hayo H. W., and Adriaan van Doorn. "A Century of Advance in Bumblebee Domestication and the Economic and Environmental Aspects of Its Commercialization for Pollination." *Apidologie* 37 (2006): 421–51.

Williams, N. M., E. E. Crone, T. H. Roulston, R. L. Minckley, L. Packer and S. G. Potts. "Ecological and Life History Traits Predict Bee Species Responses to Environmental Disturbances." *Biological Conservation,* forthcoming.

Index

Africanized bees, 9–10
  economic impact, 10
  impact on honey bees, 11–12
agave, 32
agriculture, industrialized, 7–8
alfalfa, 7, 14, 21–26
alfalfa hay, 23
alfalfa leafcutter bees, 22–25, 136, 224
alien species. *See* invasive species
allodapine bees, 99
allotype, 68
almonds, 9
andrenid, 91, 92
Andrenidae, 233
Antarctica, 3, 134
anthers, 18
aphids, 55
Apidae, 233
apoid wasps, 54–55
Arctic, 48, 50, 87
Arica, Chile, 2

arid lands, 3
Arizona, 141–43
articulated nomad bees, 103–4
asters, 92–93
Atacama Desert, Chile, 1–2, 4, 94, 206–8
Australia, 79–81

band-footed sweat bees, 93–94
Bänziger, Hans, 90
bee biodiversity, 187–88, 260–61
  patterns in, 134, 138–41, 145–50 (*see also* extinction(s))
bee-collecting devices, 135–36
bee families, 233
bee flies, 161–62
bee god, 43
bee-identification courses, 3
beekeepers, problems facing, 9
bees. *See also specific topics*
  busy/hardworking vs. lazy, 47–50

bees (*continued*)
  catching, 134–37
  environmental and ecological
    importance, 3–6, 194
  as environmental indicator,
    195, 197, 200–204
  generalist vs. specialist, 76, 91,
    144, 146–47, 201, 203, 206,
    207, 209–10
  global distribution, 134, 141
    (*see also* bee biodiversity)
  having to commute long
    distances, 21
  historical perspective on, 54, 55
  misconceptions about, 41, 42,
    45–48, 245–46
  misidentification and differenti-
    ation from other insects,
    51–59 (*see also* species
    identification)
  natural enemies of, 256–57
    (*see also specific enemies*)
  relatives of
  carnivorous, 56
  closest, 54
  size, 202 (*see also* queen bees,
    physical appearance)
  steps to help, 261–62
    (*see also* pollinators, steps to
    help preserve)
  unusual looking, 52, 53, 59
  viewed as pests, 213
  waspy ancestors of, 39–40
  worshipping, 43
bee sampling, 134–37
Bees of the World, 55–56

beetle larva, 158–61
beetles, 12, 59. *See also* Francisco
  oil beetle
bee wolf, 35–36, 169–71
bicoloured agapostemon, 103,
  104
biodiversity. *See* bee biodiversity
blueberries, 14, 18–21, 174–75
Brazilian bees, 10
British Museum of Natural History, 63. *See also* Natural History Museum
brood cells, 161–62
Buchmann, Steve, 83–84
bumble bee experts, 177
bumble bees, 44, 216
  blueberry flowers and, 21
  colonies, 45
  decline in, 180–83
  diseases, 154–55, 179, 180
  foraging, 21
  historical perspective on, 177
  hives, 47
  honey, 44
  larva, 99
  nests, 218
  pesticides and, 182, 183
  physical description, 177
  "Plight of the Bumble Bee,"
    179–80
  research on, 179–83
  seasonal variations, 206
  social behaviour, 47, 105, 112,
    177, 178, 204, 207
  species, 44, 50, 178, 180, 182,
    183 (*see also specific species*)

wasps and, 170
bumble bee wolf, 169–71
bumble bee workers, 45, 48
buttercups, 92
"buzz-pollination," 60

Calgary, Alberta, 100–103
Canada, 20, 24, 26, 48, 77, 138, 142, 149–50. *See also* Calgary; Ontario; Toronto
Cape Breton, 87
carpenter bees, 48–49, 82–83, 97, 151, 154, 217
cellophane bees, 89–90
cellular phones, 12
Chile. *See* Atacama Desert
chromosomes, 108, 126, 127, 129
Clarke, W. D., 23
climate, 142–43
climate change, 149, 183–85, 188, 209
co-evolution, 90
coffee, 7, 23, 29–32
Colla, Sheila, 180–81
Colletidae, 233
colony collapse disorder (CCD), 12 causes, 12–13
communal society, 103
communication skills, 45–47
conopids, 168–69
Costa Rica, 30, 31, 33
Cretaceous period, 54
cross-pollination, 18
"cryptic species," 72
cuckoo bees, 50, 56, 103–4, 135 eggs, 49, 97, 164–65

extinction risk, 204
gaining entry into host nests, 49, 165–66
identifying, 58
macropis, 176–77
that lay eggs in nests of only one species, 204
types of, 56–57
cuckoo bumble bees, Ashton's, 178, 180, 190

dandelion pollen, 93
Danforth, Bryan, 145, 146
Dawson's burrowing bees, 123
death of bees, 170, 171, 253–54. *See also* extinction(s); queen bees, death causes of, 162 (*see also specific causes*)
defecation, 98–100
deserts, 94, 143, 145–46, 206–7. *See also* sandpits; *specific deserts*
diet of bees, 91, 94, 95
digger wasps, 55
diploid organisms, 108, 127
disease, 12, 163, 179, 180, 225. *See also* parasites
DNA, 74, 108–9 mutations, 74 used to estimate population size, 204–5
DNA barcoding, 67, 72–77, 198 cost, 94, 198 reducing the demand for, 198–99 semi-automated, 197

Dordogne, France, 151
Dukas, Reuven, 170

ecological systems, thresholds for stability of, 189–91
ecosystem services, 191–92
egg-laying, 39, 107
eggs, 92, 93, 97–98, 111, 161
  of cuckoo bees, 164–65, 204
Ehrlich, Paul, 189
Else, George, 35–36
endangered species. See extinction(s)
energy crisis, 193
environmental indicators. See also bees, as environmental indicator
  criteria for good, 195–97
eusocial bees, 105, 106, 110, 111
eusocial colonies/societies, 104–5, 114
eusocial sweat bees, 118–19
evolution, 54, 55, 98, 163, 167. See also natural selection
  of cuckoo bee lifestyle, 49, 56
  division of labour and, 107–8
  of other insects to resemble/ mimic, 59–60
  of stings, 39, 61
evolutionary tree for bees, 145, 203, 207
excretion, 98–100
extinction(s), 175–77, 188–89, 227
  causes of, 129, 175, 185
    (see also specific causes)
  consequences of, 3, 5, 20

  risk of, 3, 130, 175–78, 185, 187–88, 204 (see also bee biodiversity; bees, as environmental indicator; human activity and bees)

"feeding the bees" ceremony, 221–22
fertilization, 108
flies
  that look like bees, 59
  wasp-like, 168–69
floral oils, 90
Florida, 117, 119
flowering plants, 143–44. See also plants
  impact of bee extinction on, 3, 5, 20
flower spiders, 168
food, human, 193
  bees and human food supply, 3, 6–7, 13
  buying organic, 219–20
  future of, 244–45
  growing, 215–16
  pollinator-dependent, 3, 5–9, 13, 31–33, 191–94, 224, 225 (see also specific foods)
food chain, 6–7
food for bees. See diet of bees
foraging, 104–5, 107, 143, 161, 202
  hazards awaiting bees while, 168–69
  pesticides and, 182–83
  for pollen, 47–48, 54, 61, 91
  suboptimal, 182–83

time spent, 47–48
foraging trips, 95–97
foreign location, bees moved to, 154. *See also* invasive species
forests, 150–51
formic acid, 12
fragrances, desire for, 124–25
Francisco oil beetle, 157–58, 161
Franklin's bumble bees, 177–78
freezing bees, 4

gardening practices, "bee-beneficial," 215–16, 218–19
gender differences in behaviour, 103. *See also* reproductive division of labour
generalist bees. *See* bees, generalist vs. specialist
genes, 126, 129
genetic variability, 205–6
   of crops, 32
Germany, 173–74
giant resin bees, 154, 186
global warming. *See* climate change
golden augochlorella, 87
golden-tailed lithurgus, 185–86
grass, walking on the, 220
greenhouse crops pollinated by bumble bees, 178–79
Grixti, Jennifer, 149–50, 208, 209
ground-nesting bees, 110, 145, 162, 175, 217–18, 220, 221

Häagen-Dazs, 214
habitat fragmentation, 184, 187, 188, 256–57

habitat heterogeneity, 140, 148–49
habitats, 3
hair on bodies of bees, branched, 57
halictid, 91–92
Halictidae, 233
hand-pollination, 30–31
haplodiploids, 129
haplodiploid sex-determination system, 108–9
haplodiploidy, 130, 205
haploid cells, 108
haploids, 129
Hebert, Paul, 73
hives, 45, 47
honey bee, 45, 47
   declining number of, 8
holotype, 68
honey, 42–43
   medicinal properties, 43
   from stingless bees, 42–43, 213, 227
honey bees. *See also specific topics*
   are not always good pollinators, 14
   colonies, 44
   complex structure of their society, 44, 47
   decline in, 242–43 (*see also* extinction(s))
   North American market flooded with imported, 13
   portion of bees that are, 44
   social lives of stingless bees vs., 106–7
   species, 8n1, 44–46
   stresses on, 12–13

honey bee wolf, 55, 171
honeydew, 55
housing with green roofs, 221
human activity and bees, 171, 175, 183–87, 257–60. *See also* *specific topics*
Hurst, Pam, 79–80
Hymenoptera, 38–39, 112

identification. *See also* species identification
of bees by their nestmates, 113
identification guides, 3–4, 69–71, 141, 226, 248
online, 197–98
user-friendly, 197–98
International Bee Course, 51
International Code of Zoological Nomenclature (ICZN), 66–67
invasive species, 152–54, 185–86
Isle of Wright, 35
Israeli acute paralysis virus (IAPV), 12

Janzen, Dan, 75–76
jewelweed bumble bees, 179

Kerr, Warwick, 9, 10
Kevan, Peter, 133, 174
killer bees. *See* Africanized bees
Knerer, Gerd, 165, 169
Kremen, Claire, 26–29

large carpenter bees, 83, 97, 151, 217
large leafcutting bees, 154

larva, 25, 89–90, 97, 98, 160–62
allodapine, 99
bee flies eating, 162, 168, 169
beetle, 158–61
bumble bee, 99
conopid, 168–69
dandelion pollen and, 93
disease and, 163
hatching and feeding, 97
honey bee, 107
life of, 98
wasp, 36, 37, 54
leafcutter bees, 84–85. *See also* alfalfa leafcutter bees
lifecycle of bees, 82
stages of, 97–100
ligated gregarious bees, 71–72
long-horned sphecodes, 165

MacDonald, John, 17
macropis cuckoo bees, 176–77
masked bees, 56
mating behaviour and strategies, 117–21, 125–26, 158–60. *See also* reproductive division of labour
Mayans, 43–44, 227–28
Megachilidae, 233
melittid, 91–92
Melittidae, 233
messenger sweat bees, 165
metabasitarsus, hind, 57
Michener, Charles, 89, 189
Middle East, 151
migration, 184. *See also* invasive species

Minckley, Robert, 141–42
mites, 11, 162
mitochondrial DNA (mtDNA), 74
mobile phones, 12
monkshood, 170–71
morphological identification, 66–68. *See also* species identification
moths, 59
  wax, 12

National Center for Ecological Analysis and Synthesis (NCEAS), 199–200
National Research Council (NRC), 8
Natural History Museum, 63–65, 68
natural selection, 59. *See also* evolution
nectar, 42, 47, 55, 89–91
nectar bars, 121–23
Neff, Jack, 95
nest defence, 104
nesting, 86–88, 106
  social bees, 80, 101
nestmates, 112–14
nests, 88–89, 101–3, 106, 218. *See also* eggs
  assistance from other organisms in constructing, 165–68
  enemy attacks on, 165–68
nest site(s), 123, 162
  artificial, 136, 137
  choice of, 201

providing, 216–19
nitrogen, liquid, 4
nomad bees, 57, 103–4
Nova Scotia, 20
nutrition, human, 32

oil crisis, 193
Ontario, Canada, 2–3, 181, 186. *See also* Toronto
orchard bees, 85
orchid bees, 124–25
organic farming, 26–29, 219–20
oviposition, 97
ovipositor, 39, 97, 98
Owen, Robin, 207–8

Packer, Laurence
  writings, 3–4
pale-footed habropoda, 158–60
pan traps, 135–37
parasites, bee, 11
  sexually transmitted, 159–61
parasitoid larva and wasps, 98. *See also* larva; parasites
paratypes, 70
pearly-banded bees, 123
pesticide-free buffer zones, 175
pesticides, 13
  avoiding the use of, 219
  impact on agriculture, 20, 22, 174
  impact on bees, 148, 174–75, 182–83, 196, 203, 204, 219, 225
pheromones, 159, 160
Pinery Provincial Park, 181

plants. *See also* flowering plants
growing bee-friendly native-
species, 215–16
Plateaux, Luc, 110
Plateaux-Quénu, Cecile, 110, 111
poey's gregarious bees, 117–19
pollen, transfer and transport of,
4–5. *See also* foraging, for pollen
pollen balls, 95–96, 103, 122,
162, 166
pollination, 3, 5–7, 17
of greenhouse crops, 178
increasing human need for, 8–9
nature of, 17
reliance on honey-bee for large-
scale, 13–14, 29
types of, 17–18, 60 (*see also*
hand-pollination)
pollination crisis, 3, 9, 190–94,
199, 213
pollination efforts, human,
13–14, 213
pollination web, 191
pollinator-friendly agricultural
practices, 213, 215, 220, 226
Pollinator Partnership, 241
pollinator research, 17, 30, 133,
200, 212, 214
pollinators
developing new, 226–27
and flower reproduction, 171
global concern about, 8, 197,
199, 200, 211–13
non-bee, 5, 7, 191
pesticides and, 174 (*see also*
pesticides)

steps to help preserve, 214–22,
261–62
various types of bees as, 14,
28–30 (*see also specific species*)
wasps and, 7, 170
pollinator species, endangered,
175–76
population size
estimating, 204–5
and genetic variation, 205
Potts, Simon, 137
Praz, Christophe, 92, 93
progressive provisioning, 99
protandry, 121
pupae, bee, 100
pupal stage, 97

queen bees, 45, 111, 206
death, 49, 106, 107, 112
mates and mating behaviour,
112, 118–21
physical appearance, 44–45,
81–82, 165–66
relation to workers, 44, 102,
104, 105, 112
"replacement," 49–50, 106
reproductive division of labour
and, 104 (*see also* reproductive division of labour)
virgin, 119, 121
Quezada-Euan, Javier, 221

rain, 3, 162. *See also* weather
rainforests, 137, 150–51, 188
Ramirez, Eric, 131–32
red osmia, 94

reproductive division of labour, 104, 107, 118, 123–26, 161. *See also* mating behaviour and strategies; poey's gregarious bees
genetic genealogies and, 108–14
Richards, Miriam, 48, 186
Ricketts, Taylor, 30–32
robber bees, 167–68
rock nettle, 91, 92
roofs, green, 221
royal jelly, 107
rusty-patched bumble bees, 178, 180–82, 190

sandpits, 147–48
Saul-Gershenz, Leslie, 157–59
Schmidt, Justin, 37–38
Schwarz, Mike, 79
self-pollination, 7, 17–18, 29
semi-deserts, 3
semisocial societies, 104–6
sex alleles, 126–30
sex determination, 126–29, 253–54
sex locus, 126–28, 130
sister-rearing, 114
Skaife, S. H., 86
small carpenter bees, 82–83, 151
social bees, 21, 48, 58, 206, 221. *See also specific species*
  environmental disturbances, 201–3
  nesting, 80, 101
  reproductive behaviour, 114, 118
  seasonal patterns, 21, 119, 201

social behaviour of bees, 3, 251–52. *See also specific topics*
socially parasitic bees, 49–50
solitary bee females, 44, 48, 103, 127
solitary bees, 47, 103, 127, 202, 203, 206
  most bees are, 44, 45, 47, 100, 103
  pollen-ball construction, 96–97
solitary bee way of mothering, 95
solitary mining bees, 57, 200, 203, 207, 218
Somananthan, Hema, 142–43
Sonora Desert, 141–42
South America, Africanized bees taken to, 9–10
specialist bees, 91–94, 144–47, 204. *See also* bees, generalist vs. specialist
species, 66–67, 187–88, 235–39. *See also specific topics*
  rarely seen, 176
species identification, 3–4, 65, 66–72, 141, 246–47. *See also* bees, misidentification and differentiation from other insects; DNA barcoding
species identification guides. *See* identification guides
sperm, 117–18
Stenotritidae, 233
stingless bee culturing, 43–44
stingless bees, 41–45, 60
  behaviour, 41, 106
  caste differentiation, 107

stingless bees (*continued*)
  colonies, 44, 105
  egg laying, 107
  food sources, 107
  vs. honey bees, 106
  honey from, 42–43, 213, 227
  Mayans and, 43–44, 227
  in rainforests, 150–51
  social lives of honey bees vs.,
    106–7
  types of, 42, 44, 89, 106, 167
  workers, 106, 107
stings and stingers, bee, 37–42,
    45, 59–61
  morphology, 60
storing bees, 4
subsocial society, 103
swarming, 106
sweat bees, 45, 93–94, 110–11,
    113, 118–19, 165. *See also*
    stingless bees

tarantula hawk, 40
taxonomy, 66–69, 77, 246–47.
    See also species identification
Thomson, James, 179
Thorp, Robin, 177
Toronto, Canada, 2–3, 151–52,
    173
tracheal mites, 11
trap nests, 136, 137
tricoloured loasa, 208–9
trident-tailed exoneurella, 81–82
type specimen, 68

United States Department of
    Agriculture (USDA), 10
urban areas, 151, 224

varroa mites, 11
Virginia carpenter bees, 48–49,
    154
vitamins, 32
vulture bees, 89. *See also* stingless
    bees

waggle dance, 46–47
"wanna-bees," 59
wasp-like bees and wasps mis-
    taken for bees, 51–53, 56–58
wasps, 39. *See also* bee wolf
  impact, 170
  larva, 36, 37, 54
  parasitic, 39
  pollinators and, 7, 170
watermelon, 26–29
wax moths, 12
weather, 142–43. *See also* rain
weeds, 186, 201, 215, 220
western bumble bees, 178–80,
    182
western domesticated honey bees.
    See honey bees
Westphal, Catrin, 136
wind-pollination, 7
wool carder bees, 152
World without Bees, 56

Xerces Society, 175–76
Xunan Cab, 44

yellow-banded bumble bees, 178, 180

yellow-footed mining bees, 147–48

York University, 3

Yucatan Peninsula, 43

Zayed, Amro, 128–30, 208, 209

zephyr sweat bees, 113